Zoophysiology and Ecology
Volume 7

Eberhard Curio

The Ethology of Predation

With 70 Figures

Springer-Verlag
Berlin Heidelberg New York 1976

Prof. Dr. Eberhard Curio

Arbeitsgruppe für Verhaltensforschung
Abteilung für Biologie, Ruhr-Universität Bochum
463 Bochum, FRG

For explanation of the cover motif see legend to Fig. 52 (p. 148)

ISBN 3-540-07720-0 Springer-Verlag Berlin Heidelberg New York
ISBN 0-387-07720-0 Springer-Verlag New York Heidelberg Berlin

Library of Congress Cataloging in Publication Data. Curio, Eberhard. The ethology of predation.
(Zoophysiology and ecology; v. 7). Bibliography: p. Includes index. 1. Predation (Biology). I. Title.
QL758.C87 591.5'3 76–10351.

Typesetting, printing and bookbinding: Konrad Triltsch, Graphischer Betrieb, Würzburg.

*This book is dedicated
to Nicolaas Tinbergen*

Contents

Chapter 2

Searching for Prey 42

Introduction

Predation is an ecological factor of almost universal importance for the biologist who aims at an understanding of the habits and structures of animals. Despite its pervasive nature opinions differ as to what predation really is. So far it has been defined only in negative terms; it is thought not to be parasitism, the other great process by which one organism harms another, nor filter-feeding, carrion-eating, or browsing. Accordingly, one could define predation as a process by which an animal spends some effort to locate a live prey and, in addition, spends another effort to mutilate or kill it. According to this usage of the word a nudibranch, for example, that feeds on hydroids would be a predator inasmuch as it needs some time to locate colonies of its prey which, after being located, scarcely demand more than eating, which differs little from browsing. From the definition just proposed consumption of the prey following its capture has been intentionally omitted. Indeed, an animal may be disposed of without being eaten. Hence the biological significance of predation may be more than to maintain nutritional homeostasis. In fact, predation may have something in common with the more direct forms of competition, a facet that will be only cursorily touched upon in this book. Hence, predation is best distinguished from other forms of foraging by only one of its consequences, in that it concludes with the mutilation or total destruction of an animal that offers some resistance against being discovered and/or being harmed. A parasite may also harm and eventually kill its host but, whereas it is in the interest of the parasite to leave its host alive long enough to complete its life cycle, such need not be the case for a mutilator, as will become clear later on. It is not all-important to draw such a borderline precisely: the mechanisms underlying the search of an ichneumonid parasite female for eggs of its lepidopteran host may be functionally similar to those of a coccinellid larva in its search for aphids. Therefore, whenever functional similarities seemed to open up new vistas, a broader approach was chosen by amalgating data on search and foraging behavior other than predation in the above sense.

The manner of presentation will be inductive rather than hypothetico-deductive; where possible, conclusions will be driven home by sheer weight of examples. The highly diverse body of facts has been so organized as to begin with the internal factors that make a predator search, or wait for prey, and recognize it. The factors responsible for the selection of particular prey individuals from a choice of many will be examined in the light of whether predators exert stabilizing selection upon their prey animals. This is followed by a review of the major hunting methods and by an examination of how particular hunting methods of a predator species relate to different types of prey or hunting situations. Special attention is devoted to the question of specialists and generalists and of how different manifestations of the generalist type of predator are affected by fluctuations of food abundance. Finally, a discussion of behavioral aspects of hunting success, and how predators benefit from communal hunting, will conclude the book. All behavior that follows seizure of prey, such as handling, eating, or the distribution of the booty among group members, if any, will be omitted purposely. The treatment as outlined above, however, will not be entirely parochial. The development of predatory behavior in the individual will be alluded to where it seems appropriate. However, because of lack of space, there will be no unifying chapter on the ontogeny of hunting that would draw together these scattered pieces of evidence.

The behavior of the prey and its defense mechanisms is alluded to whenever it is felt necessary to elucidate the predator's behavior and the question of its adaptedness.

The major significance of behavioral studies of predation is seen with regard to two particular points, the second of which will be considered in some detail: (1) Predators exert a noteworthy selection pressure on their prey species because, in general, the action in terms of natural selection is uniquely clearcut; the prey is killed or it escapes from attack. Because of this, studies of natural selection of prey species by predators bear out directly and numerically the efficacy of the anti-predator mechanisms involved. A similarly clearcut measure of the survival value of a particular predatory behavior would be difficult to obtain: a predator that has failed tries anew and it is difficult to relate its failures to its Darwinian fitness. (2) Being originally an ecological issue, predation has been studied widely in ecology and ethology. These two fields have made conspicuous advances in the recent past although almost in complete isolation from each other. There has been a pioneering and stimulating attempt to integrate ecological thinking into studies of behavior (Klopfer, 1962) but this work has been virtually outdated by the conspicuous upsurge of model building by modern ecologists. From these models ethologists could profit through inspiration for a fresh and evolution-oriented approach. Ecologists might profit by assimilating findings on behavior into their models. These models (reviews MacArthur, 1972; Emlen, 1973) make assumptions on parameters of hunting behavior, such as search time, pursuit time, selection of diet, etc. To the ethologist it is often not clear how complete this list of behavioral parameters is, how they interact with one another in reality, and in which way they are affected by the type and the density of prey, the presence of other predators, etc. (but see Holling's, 1965, 1966 "experimental component analysis"). It is here that ethological studies of hunting behavior can contribute greatly. Once the behavioral properties of a predator-prey system are more completely known, ecological models making predictions about this system will almost certainly gain in realism, precision, and perhaps, in generality. By integrating behavioral findings and ecological theory, a true synthesis, which might onerously be called "eco-ethology", could one day be achieved. To stimulate re-

2

search into linking both fields a few chapters will be concluded by brief eco-ethological excursions.

The hunting behavior of predators offers a colorful picture with countless facets, few of which, however, have been the subject of more penetrating studies. Anecdotal reports prevail. I have endeavored to draw from both sources where a fair degree of reliability could be taken for granted. Despite the multitude of aspects hunting behavior offers to the biologist, my intention was to establish generalities, regardless of the taxa of predators involved. Such an undertaking, however, meets with difficulties. As a rule, prey species have evolved defenses against many different predators. Often, a particular defense takes some general form, i.e. is tuned to more than one predator. As a consequence, predator species constantly strive to "invent" predatory novelties by the help of which they avoid the established defense mechanisms of their prey. This has led to a bewildering variety of hunting tactics, to which competition between predators and their different phyletic levels have added. Therefore, any grouping of predatory behavior patterns in a few functional categories, as will be attempted in this book, is aimed primarily at reducing chaos; at the same time it allows to fill in the details of the picture. Moreover, basic categories can be added easily, should they have been overlooked.

One may speculate whether this gross functional classification of hunting tactics reflects basic selection pressures exerted by the defense systems of prey animals. Another book is planned to pursue the fascinating problem of the co-evolution of predator and prey, dealing with the anti-predator mechanisms of prey animals. A significant step in this direction is the recent book by Edmunds (1974).

From the widely scattered literature on predatory behavior, I have selected those papers that in some way or another have contributed "key" discoveries or ideas. I apologize to all authors of good articles not included. The search for literature was largely terminated in 1973, whilst a few works have been added from 1974 and 1975.

Acknowledgments. The present book could not have been written without the help of many friends, colleagues, and institutions. My special gratitude for constructive criticism goes to those who read one or more chapters: Dr. G. M. Burghardt Chapter 3, Dr. J. R. Krebs Chapters 2, 4, and 5 B, Dr. H. Kruuk Chapter 3, Dr. J. N. M. Smith, Chapters 2, 3, and 5 B, and finally to M. Milinski who carefully examined the entire manuscript. However, any error or bias that has remained is entirely my personal responsibility. Furthermore, I enjoyed countless stimulating discussions with my pupils, M. Milinski and Dr. W. Schuler, whose hardheaded criticisms are deeply appreciated. Many other colleagues supplied me with indispensable material such as reprints or their published figures. Others allowed me to use their unpublished material and the help of R. S. O. Harding, J. Hatch, and G. Thomas is gratefully acknowledged in this respect. J. Aschoff, L. M. Dill, E. Reese, M. Robinson, and R. Royama supplied invaluable written information. Many publishers kindly granted permission to use illustrations or tables from their books or journals. Mrs. J. Herden and Mrs. H. Möbius as well as my friend Dr. Blaich produced a number of the illustrations. Mrs. U. Schwalm, Miss J. Schmale, and Miss D. Stein, and my wife Dorothea shared the tedious job of typing various drafts of the manuscript. The Deutsche Forschungsgemeinschaft helped generously by granting financial aid for several of my research projects beginning in 1960, in the course of which, findings and ideas arose that benefited the book materially. At this point I must stress that my own ideas have been profoundly influenced by the work of Niko Tinbergen and his group at Oxford, who made so many important contributions to the field covered in this book.

Finally, I would like to express my gratitude to Professor Dr. H. Langer who smoothed many editorial problems in the course of preparing this book and to the publisher Dr. Konrad Springer who generously made space available when in the course of writing it became clear that the space allotted originally proved insufficient to cover the field with some depth. Mrs. Jean von dem Bussche meticulously eradicated inconsistencies and improved my English.

There are many biologists with first hand experience of the predatory behavior of animals, who may have views on many issues of predation which differ from mine. It would be a pleasure for me to receive their comments and criticisms.

Chapter 1

Internal Factors

The phenomena presented in this chapter are, to a large part, typical of feeding behavior in general and not just of predation.

A. Hunger: Expression through Overt Behavior

Hunger among the internal factors, is the most pervasive in causing a predator to search for and hunt prey and, after successful capture, in determining the amount eaten. Hunger as an internal state eludes direct observation, and only rarely can it be observed more directly. The tentacles of hydra cultured in the laboratory lengthen steadily as time goes by since the last feeding. Tentacle length reaches a maximum after four or five days and decreases thereafter. The increase and subsequent decrease of tentacle length closely parallels the increase and subsequent decrease of hydroxy-proline as a likely measure of nematocyst content of hydra tentacles (Murdock and Murdock, 1972). Thus, tentacle length in combination with knowledge of the antecedent deprivation of food may serve as a useful indicator of hunger in hydra.

In experiments with a fish v. Holst (1948) demonstrated *visual vigilance* leading to prey capture to vary with hunger. Prey species drifted singly towards a fish from the front, and he used the dorsal light reaction as an indicator of *visual vigilance*. Fig. 1 illustrates the gradual "awakening" of visual vigilance, as expressed by the increase of body tilt towards the light, after the onset of illumination. At first, prey (worms) pass, without being fixated and as vigilance rises they are fixated and finally snapped up. Each act of fixation and each capture is accompanied by a steep increase of body tilt. These jumps of visual vigilance are prey-specific and cannot be elicited by objects other than prey. The time course of each rise of tilt, when examined closely, is the same regardless of whether the fish snapped up or merely fixated the worm. Cues from worms perceived in the mouth during capture can therefore be dismissed as causal agents of arousal; it is solely due to visual prey stimuli. The small size of the tilt changes during the first captures may indicate that overt attention leading to capture needs some "warming-up" (see also Tugendhat, 1960), as does visual awakening. It can be ruled out that the urge is hunger, since the period of darkness preceding the experiment is long as compared to that of awakening. As the fish becomes satiated the steep humps of the tilt angle curve decrease in size though the level of general visual vigilance remains. This suggests that the brief increases of tilt are due to the momentary arousal of the capture mechanism which in its turn operates as a function of hunger. This assumption is supported by still another fact. The distance by which the fish respond to approaching prey fluctuates in parallel with the sudden changes of body tilt and is greater with longer deprivation of food. In conclusion, there is a hunger-dependent "central prey excitation" which expresses itself as a short-term rise of a more stable general visual vigilance which must have a minimal level for capture to occur.

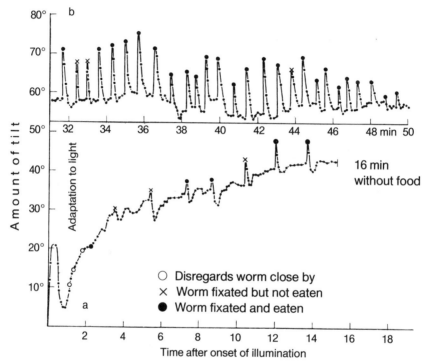

Fig. 1. (a) 'Central prey excitation' in *Pterophyllum eimecki*, as measured by its dorsal light reaction to a lateral light, after 8 h in the dark and provision of food, single *Enchytraeus* worms. (b) Food provision was resumed after 16 min without prey. Ordinate: The angle of tilt of the median plane subtended from the vertical. (After v. Holst, 1948)

The term "hunger", as used here, refers to the resultant of a whole set of messages of internal, e.g. of gastrointestinal origin (Miller, 1957; Dethier and Bodenstein, 1958; Dethier, 1967) which signal a caloric or a more specific deficit (review de Ruiter, 1963). Hunger indicates the reverse of satiety so that its value is minimal when hunger is maximal, and conversely. Even though the internal signals are not directly measurable, this definition can be used operationally, because the state of the signals varies with the feeding schedule, quantity and quality of food, and the metabolic conditions imposed on the subject. Other things being equal, the hunger state uniquely determines the predator's tendency to respond to the prey situation. By contrast, the readiness to feed will not always define the hunger state. (1) Responsiveness drops to zero as soon as satiety extends beyond a certain level. (2) Responsiveness may reach a plateau beyond which further increase of hunger has no effect. For these and other reasons, different measures of hunger will often not concur (Hinde, 1959; Beukema, 1968). It therefore seems advisable to define hunger in terms of the previous feeding schedule and metabolic conditions, rather than in terms of overt feeding behavior (de Ruiter, 1967).

I. Predatory Schedules

1. Patterns of Satiation

In a most thorough study of feeding behavior Beukema (1968) deprived three-spined sticklebacks *Gasterosteus aculeatus* L. of food for varying times. Food intake reflected the length of the deprivation period (Fig. 2). The longer deprivations caused a doubling in the rate of eating in the first hour, although all fish began eating with empty stomachs on the day of their test. This means that hunger is determined by at least two components, i.e.

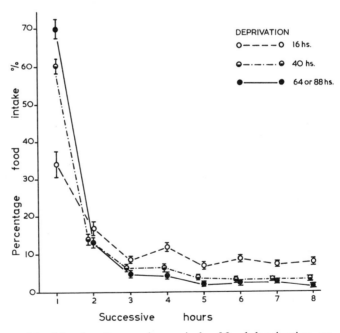

Fig. 2. The daily pattern of food intake after varying periods of food deprivation, expressed as the percentage of 8 h intake by three-spined sticklebacks. Each point represents an average (±s.e.) of 24 observations, i.e. 3 on each of 8 fishes used. (From Beukema, 1968)

a gastric and a "systemic" component dependent upon the interval of fasting imposed on the subject. The amount eaten during the second hour is equal after all three deprivations. The proportions of the later hours are smaller as deprivation ran longer (see also Precht and Freytag, 1958). This results in all fish eating approx. 12% of their body weight on the day of testing, irrespective of the length of the deprivation period.

The slackened eating after the first hour by subjects deprived for longer, may be a consequence of their voracious feeding during the first hour, rather than of the deprivation level. After a given deprivation individual sticklebacks that ate more than average during the first hour, took a lower than average amount during the last seven hours of the daily feeding period. Thus, at the same deprivation, the amount eaten during the first hour controls to some extent the quantity eaten later.

This pattern of rapidly declining food intake after starvation is obviously widespread and was also found by Salt (1961) in *Amoeba proteus* feeding on *Tetrahymena pyriformis*, a ciliate.

After satiation the stickleback continues feeding intermittently if the prey situation allows. It is never satiated for a long period of time, as is for instance a mammal which intersperses conspicuous "digestive periods" between bouts of feeding (Fig. 3). The feeding patterns of fish and a praying mantis (*Hierodula crassa*) on one hand, and of mammals on the other, are but two extremes of a whole spectrum of intergradations, if appropriate allowance is made for length of pauses and feedings.

The few studies of how hunger affects feeding responses in amphibians and reptiles have yielded rather unexpected results. Strangely, hunger does not appear to influence feeding in American toads (*Bufo fowleri*) since the amount of food consumed is independent, within wide limits, of the length

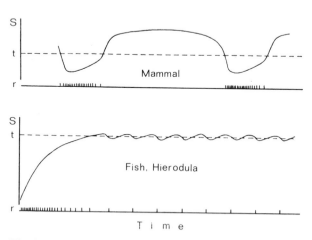

Fig. 3. Hypothetical diagram of time course, as suggested by behavioral data, of overall satiation S, resulting from the various satiety signals in mammals, fish, and a praying mantis. t = satiety threshold above which no feeding responses occur; r = their time pattern. (Combined from de Ruiter, 1967; Holling, 1966)

of time without food (Heatwole and Heatwole, 1968). Although the feeding behavior of rat snakes (*Elaphe obsoleta*) reminds one of the mammalian pattern of satiation (Fig. 3), puzzling points remain. This snake preys and feeds on mice only at intervals of several days, even though they are offered every day; its average meal comprises a total of 3 to 4 mice but can reach 19. It seems as if hunger must rise above a threshold level which is not reduced by just one mouse. Before shedding the skin, intervals between successive meals lengthen. It is strange that after mice had been withheld for six weeks, feeding on mice became irregular or ceased altogether; none of the four snakes displayed a compensatory elevation of food intake after deprivation. The reason why these snakes at times exhibit voracious feeding and at other times fast for long periods may hopefully be elucidated by a more extended study of free feeding (Myer and Kowell, 1971).

An unexplained temporary fasting has also been found in some avian (Scherzinger, 1970) and canid predators (Mech, 1970). In the majority of cases predators fast voluntarily during certain phases of their life. The systemic relations to hunger are poorly understood. Fasting occurs during periods of mating (H. and J. van Lawick-Goodall, 1970, p. 181; Schaller, 1972, p. 141), pregnancy (pers. obs.), pseudopregnancy (de Ruiter, 1967) and gestation (Keenlyne, 1972) in mammals; during parental care for eggs (Lack, 1954, p. 139) and pups (Irving, 1972, p. 81) in birds and mammals, respectively; also prior to skin-shedding in snakes (see example above).

After unphysiologically long, enforced starvation as reported by Ivlev (1961) for pike (*Esox lucius*), tench (*Tinca tinca*), and bleak, (*Alburnus alburnus*) food intake may cease altogether. Similarly, Windell (1966) noted that the appetite of bluegill sunfish (*Lepomis gibbosus*) increased to a maximum after four days' starvation and then decreased, accompanied by degenerative changes in the pyloric caeca by the tenth day. In principle, appetite could, after long fasting, die off because of pathological changes incurred by the digestive system or because of an "atrophy" of the readiness to feed, due to lack of use (e.g. Heiligenberg, 1963: *Pelmatochromis subocellatus kribensis,* Cichlidae). The above observation on sunfish, however, bears out clearly that extraneural events may be sufficient to explain starvation-related disturbance of the feeding response system.

A pattern of satiation for which gastric factors alone are a sufficient explanation was demonstrated by Sandness and McMurtry (1972) in a predatory mite (*Amblyseius largoensis*). When permitted free feeding on a herbivorous mite (*Oligonychus punicae*) confined in small arenas on excised avocado leaves (*Persea indica*) the predators spent increasingly less time in feeding on successive prey objects (Fig. 4). A peak in the length of the digestive pause(s) followed the first peak in duration of feeding while it preceded later ones. Therefore the length of the digestive pause cannot be the sole causal determinant of the cycling of feeding duration; however gastric factors could, as indicated by two findings: (1) The time at which the second prey was fed and the duration of feeding was similar regardless of the preceding starvation period. (2) Voluntary digestive pauses, i.e. those in Fig. 4, peaked prior to peaks of feeding duration.

The fact that the mite predator killed more than it consumed shows that the readiness to kill is not determined by hunger alone (see also Chap. 1. D). Surplus killing may have been due to excessive prey density, as in mammalian carnivores, where it results from overstimulation of super-abundant and vulnerable prey (Kruuk, 1972 a).

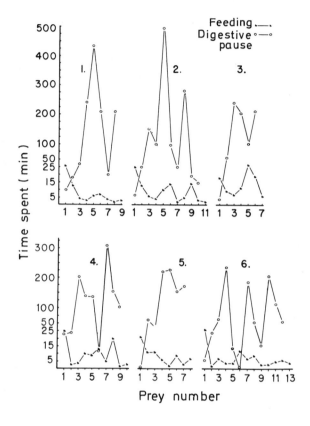

Fig. 4. Comparison of time spent feeding on successive prey (*Oligonychus punicae*) with length of digestive pause after each meal for six predatory mites (*Amblyseius largoensis*) during a 24-h period. (From Sandness and McMurtry, 1972)

2. Feast and Famine

Under laboratory conditions food supply is usually held constant and prey density high (e.g. Beukema, 1968; Holling, 1966). The question arises as to whether the rules thus observed also prevail in the wild. Though the conditions mentioned are not often met with in the wild, predators when presented with unlimited prey gorge more than is needed for growth and metabolism. Gluttonous feeding leads to flesh being excreted undigested by carnivorous fish (Winberg, 1960, quoted by Beukema, 1968) and mink (*Mustela vison*) (Errington, 1967), for example. Further, given the opportunity large carnivores gorge far beyond their subsistence diet; spotted hyenas (*Crocuta crocuta*) by a factor of up to seven (Kruuk, 1972 b, p. 75 seq.) and lions (*Panthera leo*) by a factor of four to five (Schaller, 1972, p. 276; see also Mech, 1970, p. 182 seq.: wolf). Predatory fish (*Sebastodes dimidiatus*) en-

10

gorge when swamped with easy prey (*Engraulis mordax*) in such a manner that one of these can get jammed in the mouth for up to two and a half hours, until digestion makes way for the next (Clarke *et. al.,* 1967; see also Lüling, 1973). A "Schlaraffenland" situation such as this reminds one of aquatic filter feeders or browsers.

The discrepancy between the maximum meal size possible, in the carnivores mentioned and others, and the average meal size, is almost certainly a result of their adaptation to famine, when intense search for prey ends in no capture for long spells of time. A hyena followed by Kruuk (1972 b, p. 76) without interruption ate at intervals of at least 2, 2, 1, 5, and 2 days. Wolves (*Canis lupus*) are able to starve for at least 17 days while still travelling in search of vulnerable prey (Mech, 1970, p. 182 seq.), and lions obtain food every 3 to 3.5 days. A nomad male followed through nine consecutive days ate during 7 of these, ingesting an estimated average of 9 to 10 kg of meat per day; another nomad male ate during only 7 out of 21 days, ingesting an estimated 6 to 7 kg per day. At certain times of the year, particularly when wildebeest (*Connochaetus taurinus*) migrate through areas of lion prides, they kill and eat almost twice as much as they need; at other times, they just fulfil their daily requirements, or are unable to meet them which results in heavy cub mortality (Schaller, 1972 b, p. 279 seq.).

From these remarks it follows that feast and famine are general attributes of the life of the large mammalian carnivores. It appears that their best strategy in respect of nutritional homeostasis is to gorge when the opportunity arises, to tide over spells of famine. If this is correct a meal above average size is a nutritional advantage that outweighs disposal of undigested meat mentioned above.

Gorging with food has still another aspect in that it may impair locomotor ability. One would expect small predators that may themselves fall prey to larger ones to eat no more than necessary, so as to permit unimpeded flight if attack is imminent. In fact, vampire bats (*Desmodus rotundus*) rarely gorge with blood to such an extent that they cannot take wing (Greenhall *et. al.,* 1971), while a lion or a griffon vulture (*Gyps fulvus*) (C. König, pers. comm.) appears not to take this into consideration.

II. Hunger and Diel Rhythms

The way in which gastrointestinal factors interact with diel rhythmicity has been rarely studied. For a praying mantis Holling (1966) found both gut content as well as time of day are determinants of the capture rate; with a given gut content the mantid will capture more flies per unit time in the morning than later in the day. Similarly, large carnivores, e.g. African wild dog (*Lycaon pictus*) and lion, do their hunting around dusk and dawn. If they have been unsuccessful during either period they will rest until the next period has come when they will try anew (Kühme, 1965). Hence diel rhythmicity dominates over hunger. It remains unknown whether this effect is due to a suppression of hunger or of behavior patterns associated with hunting.

Mueller (1973) in a study with two hawk species, found number of prey killed and amount of killed prey eaten to vary both as a function of deprivation time and of time of day. The situation is more complex with the caching and retrieving of surplus food by one of the species, the American kestrel (*Falco sparverius*) (Mueller, 1974). Whilst caching varies with deprivation time and not with diel rhythm, the retrieval of previously cached food clearly varies with the time of day, though a less clear influence from hunger is also recognizable.

III. The Ramification of Hunger Effects

The rate of prey capture is only one of the many signs by which the degree of hunger finds its expression. Subtler signs become obvious if one examines in detail the stimuli necessary to elicit attack and the sequence of events finally leading to capture and ingestion.

1. Capture-eliciting Prey Stimuli

As a rule, hunger tends to increase the range of objects recognized or accepted as food. Sticklebacks fixate or snap up inedible small objects resembling prey progressively with increasing hunger (Beukema, 1968). *Pterophyllum eimecki* also fixates such objects at high hunger levels but its "central prey excitation" at the moment of fixation is conspicuously lower than when dealing with real prey, as measured by its dorsal light reaction (v. Holst, 1948, see also p. 6). As satiation grows, "central prey excitation", triggered by both inedible as well as real prey objects decreases in parallel, thus indicating that both classes of objects in some way arouse the capture-releasing mechanism though to different degrees.

Using a different method, Swynnerton (1919) explored the relationship between palatability of prey and satiation. Mammals, reptiles, and birds were offered one insect at a time from an array of more than thirty different species. When hungry, the predators accepted them indiscriminately but, with growing satiation, only the most palatable. Similarly, kingfishers (*Alcedo atthis*) when starved, attack fish that are larger and that swim at a greater depth than they normally would (Kniprath, 1969). Famine due to heavy snow cover may turn insect-eating songbirds into fierce marauders killing others for food (Roth, 1971).

If certain digger wasps (Scoliidae) cannot obtain spiders of the genus *Epeira*, their speciality, they resort to other spider genera to feed their brood (Fabre, 1879, quoted by Edwards, 1963).

In *Epiblemum* spiders (Salticidae), black, round, flat dummies eliciting capture movements are tolerated as larger than normal prey with growing food deprivation, until finally dummy size extends into a range which in normal spiders releases flight responses (Drees, 1952; see also Hoppenheit, 1964 a: *Aeschna cyanea* larvae). In European toads (*Bufo bufo*) prey capture grades into fleeing as a function of increasing size of a black figure on a

12

light background (Ewert, 1968, 1970). As American toads become satiated their upper limit of prey size decreases, whereas the lower limit remains the same (Heatwole and Heatwole, 1968). Thus, as in *Epiblemum* und *Aeschna*, hunger extends the size range of capture eliciting objects into the range of the flight response to potentially dangerous objects. Hence, hunger-motivated awareness and approach responses remain sensitive to a potential danger only when hunger is near its peak.

2. Search Behavior

Searching for prey is "any hunger-dependent behaviour of a predator likely to bring a prey within range of its exteroceptors" (de Ruiter, 1967, p. 100). Methods of searching will be more fully discussed below (see Chap. 2.).

Search for prey takes varying amounts of time. For example, with oystercatchers *Haematopus ostralegus* from 15% (Drinnan, 1957), 17% and 20% in two protozoa, respectively (Salt, 1967), to 37% in the predatory whelk *Thais lapillus*, which searches and feeds during two consecutive tides and then goes into a digestive pause of the same duration (Connell, 1961). Observations in the wild indicate that the percentage of total time available devoted to feeding behavior varies inversely with the abundance of food. During food scarcity this time may shoot up to almost 80% (Gibb's data in Lack, 1954, p. 138) in goldcrests and tits, and wolves tend to travel longer distances if the need arises. A pack of 15 to 16 wolves rested less than a pack of five, presumably because the larger pack had to expend more energy in hunting to find enough food to feed all its members (Mech, 1970; Mech *et al.*, 1971 in Mech and Frenzel, 1971). Lions of the East African plains behave differently. During the lean dry season, when cubs are dying of starvation, lions are no more active than at other times of the year: "they were obviously sparing in their expenditure of energy." (Schaller, 1972, p. 122). A predatory mite combines both methods of coping with starvation. When searching for its prey, a herbivorous mite, the predator first increases its speed of movement with the level of starvation, but decreases it at higher levels up to the point of complete inactivity or death (Sandness and McMurtry, 1972). Since phytoseiid mites do not discover prey by vision, an increase or decrease in the rate of movement would have a similar effect, such as an increase or decrease in the size of the perceptual field of a visually hunting predator like a mantid (p. 15).

In still another way coccinellid larvae (*Adalia bipunctata, Coccinella septempunctata, Propylea quatuordecoinpunctata*) feeding on aphids (*Aphis fabae*) enhance their chances of finding prey during food shortage. They search along the nerves of the leaf or the leaf edge, i.e. move more or less in a straight line. Immediately after eating an aphid, they turn on the spot and thereby increase their chance to find more prey which typically form clusters (Banks, 1957). Significantly, Dixon (1959) found that the turning rate of coccinellid larvae declines with increasing hunger, so that the straight-ahead movement leads these predators eventually to new clusters of aphids.

Numerous studies with laboratory rats have also shown that running activity increases as the subjects become hungrier. Hinde (1970, p. 259 seq.) critically reviewed the evidence and discussed the question of whether food deprivation affects activity per se or increases responsiveness to stimulation. From experiments with environments that had been impoverished to various degrees, it became evident that there is some increase in a constant environment due to hunger but disproportionately more when the environment is varied. Further work showed that complex stimulation per se does not enhance activity; instead, stimuli previously associated with food are particularly effective. Such a result is not surprising since selective responsiveness to external stimuli is often the most important feature by which different types of appetitive behavior, of which food-seeking is only one, can be categorized. Moreover, the conclusion is in line with the finding reported previously that hunger enhances awareness of potential prey objects in predators as different as arthropods and vertebrates.

An animal roaming over its home range is not necessarily hungry and may display "exploratory behavior" that serves to make the animal intimately acquainted with the topography. Exploratory behavior is thought to be instigated by a motivation in its own right that becomes reduced by perception of stimuli from objects or situations after a certain time of exposure to them. However, even from this point of view, the animal's roaming over its territory must be in part determined by hunger in the animal for it is enhanced by it; the animal's movements are then largely restricted to areas where it had previously found food; furthermore, the approach responses observed are directed primarily to objects signalling food (see Chap. 2. B.).

IV. The Motivation Underlying Feeding Responses

1. Hunger Thresholds of Feeding Response Components

In general, feeding responses appear to be subservient to nutritional homeostasis, one sign of which is the notable stability of body weights. However, a number of phenomena is not readily explained by invoking hunger as the only motivation underlying feeding responses. An analysis of the completeness of feeding responses as a function of hunger may lead to an understanding of the functional organization underlying acts of predation.

The Distance of Reaction to Prey. In a complete feeding response the components search or orient, pursue and/or stalk, capture, and eat, are following in that order. The detection of a prey by a praying mantis may lead to the full predatory sequence consisting of awareness, indicated by a movement of the head, stalk or pursuit, and strike. A fly moving in front of the mantis elicits awareness at all times, even directly after satiation. The animal then fixates the prey when it is within a range of 4 cm and with rising hunger finally orients to flies, up to a distance of 17 cm. With a prey showing up anywhere in the field of vision the maximum distance of awareness decreases as the angle subtending between the prey and the median plane increases (Fig. 5); at the same time the reactive field of awareness increases with growing hunger (see also p. 6). The elongate shape of the reactive field relates to the morphological properties of binocular vision. In predators locating prey by odor or tactile stimuli the reactive field would be expected to be roughly circular.

By comparison, stalking or striking a fly cannot be elicited earlier than eight hours after satiation. With rising hunger the maximum distance of these response components increases curvilinearly up to a maximum levelling off at approx. 7 cm, i. e. far below the curve of the awareness response (see above).

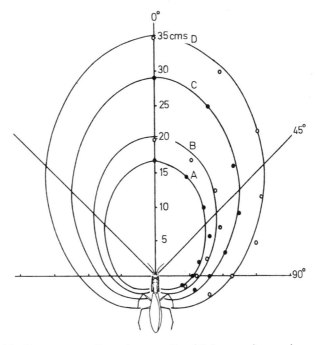

Fig. 5. Shape of reactive field of awareness of praying mantis with hunger increasing from low *A* through *D* (average of eight replicates). Distances measured along seven radii from 0° – 108°. (From Holling, 1966)

The absence of a hunger threshold of the awareness response, as opposed to stalking and striking, was interpreted by Holling (1966, p. 23) to be adaptive, "for while it is only necessary that hungry predators be aware of prey, it is extremely important that both hungry and satiated predators be aware of potentially dangerous objects." Significantly, another mantid (*Stagmatoptera biocellata*) responds with the awareness response at a greater distance than with a defensive display, as elicited with a moving songbird (Maldonado *et. al.*, 1970). The threshold difference could mean that the mantid displays its defense only as a last resort, i. e. when it has become more likely that it has been detected by a nearby predator.

15

Elicitation of the predatory sequence of Holling's mantids always ended up with ingestion of prey. The question of whether hunting per se or hunting plus eating leads to a drop of the awareness distance, was answered by Drees (1952) using a salticid spider (*Epiblemum scenicum*). It responds to prey dummies at ever-increasing distances, up to 10 cm after confinement in the dark from between 4 to 15 days. With repeated elicitation of turning towards and finally jumping upon the prey dummy, the whole sequence waned, as did the maximum reactive distance. Thus it appears that the performance of response alone can bring about contraction of the reactive field, for the increase of hunger must have been negligible during testing, which was short in comparison to the period of confinement prior to testing. Like mantids, the spiders responded with awareness to maximum distance stimulation after shorter periods of deprivation than those responsible for the full hunting sequence, and the same teleonomic explanation is applicable.

Feeding Components Following Orientation to Prey. For all components that follow orientation hunger may be necessary though varying widely among taxa (Fig. 6). In a salticid spider (*Phidippus clarus*) fixation of a prey with the two central pairs of eyes appears at the lowest hunger level for any feeding component. Gardner (1964) found that post-orientation components of hunting required more appetite but she could find no differences in hunger level between them. A very similar organization holds in another salticid, *E. scenicum,* but here Drees (1952) could experimentally separate the conditions also for these components appearing later. The hunt begins with a rapid pursuit followed by slow stalking if the prey object is moving slowly. However, if it moves quickly stalking may drop out and the spider tries to over-power the prey by a rapid pursuit alone. Moreover, in habituation experiments in which the spider was offered a prey dummy many times in succession, Drees could differentiate betwen the pursuit and the final jump

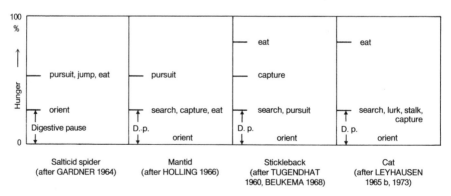

Fig. 6. Relation of the various hunger thresholds of feeding response components in four different predators. Absolute threshold levels chosen arbitrarily. At zero eating stops. 'Search' in the mantid is restricted to awareness of moving objects. (Modified and extended from Holling, 1966)

16

in still another way. When habituated to a quickly moving dummy from a distance of 8 cm the jump was not performed before the pursuit, whereas when habituated by a slowly moving dummy at 4-cm distance both components were omitted simultaneously.

As illustrated by Figure 6, a mantid requires more hunger for the active pursuit of a fly than for all other feeding components. This is what, in functional terms, would be expected from an ambush predator as compared to one that actively searches for prey.

A stickleback fixates potential prey much more often than it grasps it, and only at high hunger levels will it eat it. In any event, the probability that it eats a grasped prey is much higher than that it grasps a fixated one. This behavior is economic since fixating requires certainly much less effort than approach and capture (Tugendhat, 1960; Beukema, 1968; see also v. Holst, 1948).

The mantid differs from all other predators mentioned in this section, in that it has no hunger threshold of the orientation component. If the functional interpretation given above in terms of danger recognition holds, one would have to assume that the stricter control of orientation by hunger in the other predators is permitted by a more effective recognition of dangers. Alternatively, the other predators would have to be regarded as less vulnerable to predators than the mantid; this appears highly unlikely for salticids and sticklebacks.

From this brief account it is clear that the sequentially ordered components of the feeding response depend upon hunger and stimuli from the prey, though in a mantis, orientation is entirely independent from hunger. Mammalian carnivores display a still greater flexibility in the chaining of motor acts leading to ingestion. As with the predators referred to so far eating requires relatively high levels of hunger, domestic cats included. But a cat may still search and lurk for prey in front of a hole after being satiated. Thus hunger and eating affect these components differently. Moreover, the components of the complete sequence, e.g. lurking, catching, tossing, killing and eating a mouse, are thought to be governed by motivations of their own rather than by a unitary predatory drive; this is indicated by different times of recovery after exhaustion of those components (Leyhausen, 1965 a, 1973). A quantitative examination of this far-reaching assumption appears desirable.

Other carnivores such as wolf, red fox *Vulpes vulpes,* spotted hyena, and lion also capture and kill prey when already gorged but will then not search for more. In nature, the conditions conducive to such "surplus killing" which has been recorded for a number of carnivores are quite restricted; it occurs on rare occasions when a predator is swamped with many prey animals in highly vulnerable conditions (Kruuk, 1972 a). Therefore "surplus killing" will not upset a proper predator-prey balance (Mech and Frenzel, 1971; Kruuk, 1964, 1972 b, p. 204; Schaller, 1972, p. 251). Hunger-independent surplus killing has also been recorded for larvae of a mosquito which when swamped with prey larvae (*Aedes aegypti*) will kill more than they can eat. Unlike surplus killing by the carnivores mentioned, this behavior may be adaptive: presumably it protects the defenseless predator pupa from being eaten by their prey, the larvae of other mosquito species, as surplus killing increases towards pupation (Trpis, 1972).

2. The Complexity of Predatory Motivation

The killing of mice by Norway rats (*Epimys norvegicus*) reinforces the view of a multitude of causal factors underlying predation. Their behavior also suggests that the killing of animals other than their own species might play a role in coping with interspecific competition. As reported by Karli (1956) the incidence of killing is 70% for wild rats and 12% for laboratory rats, with no difference between the sexes. Relationships between the tendency to kill and variables related to feeding are complex. Hunger, and probably the fact that rats inexperienced in killing feed on dead mice facilitate the first mouse-killing. However, hunger is not necessary to maintain mouse-killing (Paul and Posner, 1973).

That hunger may be the major incentive to kill is also indicated by Paul's (1972) observation that, in general, rats kill a mouse more readily when hungry, and those that have killed it are more prone to eat it, than others which have not. However, most kill-experienced rats which kill when hungry continue to kill when satiated, as reported above for carnivores. Thus again, initiation of serial killing may be hunger-dependent but maintenance may be a matter of opportunity. Killing experience and hunger are related in a complex way: with growing experience the killing response becomes increasingly detached from hunger motivation (Paul, 1972). This is in accord with the fact that pharmacological treatment affects killing and feeding differentially. The evidence supports the notion that they are two separate behaviors (Krames *et. al.*, 1973).

The motivation underlying the initiation of killing may be a specific hunger, especially if one recalls that rodents are genuine omnivorous predators rather than herbivores (Landry, 1970; Thomas, 1969, 1971 a, b, c). Even so, there is a strong suggestion that mice are not merely attacked for food. Whereas familiarity with a particular food increases the probability of eating (Chap. 3. F. II.), familiarity with live mice decreases the probability of attack by rats (Paul, 1972).

Observations reported in this section raise the question as to whether predatory behavior also occurs normally without hunger. An answer comes from studies of the ontogeny of food-seeking behavior. The intensity of young European flycatchers (*Ficedula* sp.) in begging for food from their parents seems to reflect hunger remarkably well since begging becomes more intense the longer the time lapse since the last feed and the less bulky the last meal has been. Also quality of food exerts an influence (Curio, unpublished).

The transition from dependent to independent feeding takes place when hunger as expressed by begging is minimal. Right after satiation the fledged young of a number of songbirds (e.g. *Ficedula, Parus*) hop around and begin pecking at diverse small particles in their surroundings. If these are edible they are manipulated for some time, decreasingly with repetition, and are then swallowed. In this way young birds learn to direct their feeding responses to edible objects (see also Hinde, 1953; Lind, 1965; own observations). At what time this exploratory pecking after satiation comes under the control of the hunger system has been elucidated by Hogan (1971) in domestic chicks only. After hatching, chicks simply peck at novel objects like grains and sand. Releasing stimuli of objects pecked and eventually swallowed become associated with some tactual feedback or manipulability they provide. As the yolk sac is absorbed during the first 3 to 4 days two changes are thought to take place: (1) a taste mechanism is switched on which increases the incentive value of food; (2) (hunger) drive reduction begins to play a role in controlling the pecking at food after periods of deprivation. When chicks provided with food are sated they learn to peck at it due to drive reduction, by generalizing from previous situations with food to earlier ones when they were hungry. This generaliziation takes place only during a critical period; a few weeks later the chicks begin to discriminate between tests after deprivation and those after no deprivation and thus, no longer generalize from the former to the latter situation. In conclusion: a feeding response, i.e. pecking at potential food objects, is first controlled by a motivation in its own right and only after that will it become controlled (in addition?) by the hunger system (see also Leyhausen, 1965 a, for cats). The conclusion is clearly compatible with the observations on passerines reported above.

The way in which feeding components become detached from hunger differs dramatically with the species involved. Whilst in chicks, feeding occurs at first independent of hunger, mouse-killing by rats takes the reverse route: hunger facilitates the first incidence while later on other motivational factors control mouse-killing (see above).

In which way fighting over food is regulated by hunger has been aptly summarized by Hinde (1970, p. 341 seq.). In a hermit crab and in many fish, fighting is enhanced with length of deprivation of food. In songbirds fighting over food Hinde writes, "is greatest in midwinter – that is when food is probably least abundant, the need for it geratest, and the time available for finding it least" (Hinde, 1953, p. 209). Observations on various songbird species are best understood when assuming that hunger does not affect aggressiveness directly but by activating searching which in turn augments the probability or duration of social encounters. It is therefore likely that, in the feeding flock, attacking a conspecific which has come too close is different from territorial aggressiveness.

Similarly, lions kill other large cats, hyenas, wild dogs, jackals (*Canis* sp.), and people, but in a manner indistinguishable from fighting with or killing a conspecific; while their repertory of facial expression is the same in all, and restricted to these contexts the face remains entirely unmoved when capturing and killing normal ungulate prey. Another similarity between agonistic behavior and killing their most important competitors, lies in the fact that lions only rarely eat their victims, be they other carnivores or lions, including those that have died from starvation (Schaller, 1972, p. 220, H.

and J. van Lawick-Goodall, 1970, p. 184 seq.). Thus, the motivation to kill other large carnivores may serve to mitigate competition over food, a functional explanation also applicable to mouse-killing by rats (see also p. 18).

V. The Diversity of Foraging Tactics

Woodpeckers employ a number of diverse movements when foraging, namely excavating, probing, locomotion, and comfort behavior. When hungry, captive Syrian woodpeckers (*Dendrocopos syriacus*) excavate less than when sated. They thereby activate the four behavior patterns more evenly and increase the diversity of their foraging tactics. By contrast, when sated, probing will be more frequent than excavating and predominates to such an extent that overall diversity is smaller (Winkler, 1972). To what extent this change of foraging relates to hunting success remains unknown (see also p. 27). How the finding fits in the 'optimal foraging model' of MacArthur (1972) will be discussed more fully below (Chap. 5. B. III. 5).

VI. Feeding Components Affected and not Affected by Hunger

Apart from some feeding components in some predators (see above) there are other properties of hunting that are not affected by hunger. In the praying mantis, studied thoroughly by Holling (1966), in salticid spiders (Drees, 1952), and cats (Leyhausen, 1973), velocity in pursuit of prey belongs in this category. A stealthy approach involves a delicate balance between two opposing tendencies, i. e. slow stalking and rapid pursuit. Hunger must not be too great if the rate of success is to remain optimal. Significantly, capture success in the mantis remains constant regardless of hunger (Holling, 1966). Similarly, the time taken by toads after orientation towards prey until they strike at it, varies regardless of hunger (Heatwole and Heatwole, 1968). By contrast, water is jetted towards prey above the water surface by *Colisa lalia* (Anabantidae) with more success the hungrier the fish is (Lüling, 1972).

With eating too, there is no easy generalization. In the mantis mentioned and in sticklebacks it takes always the same amount of time with the same prey, irrespective of hunger levels (Holling, 1966, Beukema 1968). In European flycatchers (*Ficedula hypoleuca, F. parva, Muscicapa striata*) battering sizeable prey (mealworms, caterpillars) against the perch decreases with growing satiation and so does handling time. Yet the properties of the prey contribute to the variation observed (Curio, unpublished).

It goes without saying that mammals eat much more avidly when hungry than when not. In many species rapid consumption of a meal is a prerogative to secure sufficient amounts if competition from stronger species is imminent as illustrated by the "biological rank order" around animal carcasses in the African plains where vultures, jackals, hyenas, and lions, beside others, try to get their share (see p. 160). The commotion around a kill, when communal hunters like hyenas (H. and J. van Lawick-Goodall, 1970,

20

p. 165) or animals such as sharks (Lineaweaver and Backus, 1970) are assembling lead predators to indulge in a "feeding frenzy" (see also Allen, 1920 a) so that conspecifics may be seriously wounded by bites intended to hit the actual victim.

Feeding may also be dangerous because it cuts off danger signals. Hedgehogs (*Erinaceus europaeus*) intent upon robbing eggs or young in gull colonies crouch in response to the stoops of the gulls diving at them; once engaged in feeding they seem to ignore the attacks (Kruuk, 1964, p. 105).

B. The Control of Feeding Responses by Factors Other than Hunger

I. The Readiness to Hunt

Apart from hunger, other motivational factors contribute to the readiness to hunt. In general these factors permit a wide range of hunting episodes per unit time to take place. This behavioral potential exceeds, as a rule, what would be necessary to satisfy the food requirements of the individual. Thus there is little room for the idea of a "fixed action potential" for predatory behavior. Evidence came from the notable occurrence of surplus killing, from hunger-independent eating ('gorging'), and other sources (Chap. 1. A. II.). Furthermore, hunting behavior of the large mammalian carnivores is adjustable to the need of the social community in the extreme. For instance, individual spotted hyenas indulge more than others in killing the quarry on which all members of the clan feed. Such a division of labor benefits fat baggy females and otherwise unfit clan members that cannot hunt down a quarry on their own, not to speak of the care of the cubs (Kruuk, 1972 b, p. 165). Furthermore, male lions of a pride do only 10 to 15% of the actual killing which is then largely shared between the pride's lionesses. However, nomadic males have to kill regularly for themselves as they have no other choice (Schaller, 1972).

The opposite extreme of such a flexible organization needs to be carefully sought for in some birds. Though not yet fully documented a fixed readiness to feed rather than hunger seems to determine the rate of ingestion by hand-reared chicks of loons (*Gavia arctica, G. stellata*). Sjölander (pers. comm., 1971) claims that they take varying amounts of fish food depending upon the size of the pieces fed to them. When fed pieces of fish or smaller fish than their parents provide, chicks starve to death for they do not compensate for size of prey item by eating more, except within rather narrow limits. Similar situations are believed to exist in another fish-eating bird (*Chlidonias leucopareius*) (Mendelssohn pers. comm., 1971). To document any one case of ingestion not compensating for meal size would be worth an experimental *tour de force*.

Another line of evidence for the existence of motivating factors other than hunger comes from fluctuations in the readiness to hunt which cannot be explained in terms of hunger. Owing to lack of appropriate stimulation 'vacuum' feeding responses occur. Various anurans (Freisling, 1948) and insect-eating birds have in captivity been observed to go through the whole sequence of insect hunting, from watching the imaginary prey to catching, killing, and swallowing, without any obvious stimulus (refs. in Tinbergen, 1952, p. 62). Furthermore, Leyhausen (1973) observed that domestic cats may chase, zigzagging over a clean floor while pawing an imaginary mouse. This latter case is particularly illuminating since the objection had been raised that "vacuum" activities may merely prove an extreme lowering of the threshold of the response; a dust particle, for example, might have been taken for an insect by the hunter. Obviously, the pattern of the chase followed by the cat suggests that it "halucinated" a mouse fleeing in a corresponding pattern; clearly the path of a dust particle cannot have been as complicated. Moreover, a pure vacuum activity in the absence of an external stimulus grades imperceptibly into a situation with the minimal stimulus necessary to elicit the response so that the distinction between the two situations may be of academic interest only.

Hunger levels have not been controlled in observations such as these. The possible objection that hunger is an alternative explanation for vacuum hunting can, however, be dismissed if the components concerned belong only to the handling of prey but not the eating of it. Koenig (1951) observed that European bee-eaters (*Merops apiaster*) batter large insects such as dragonflies against the perch before swallowing them. These same movements are also performed in the absence of any prey in the beak. Strangely, the bee-eaters did not use mealworms, offered in abundance, to perform these movements; instead they swallowed this type of prey intact. By contrast, bees, wasps, and grasshoppers were invariably 'killed' this way, even if offered dead.

Captive raccoons (*Procyon lotor*) indulge in "washing" food objects in water though there is no need for it. The "washing" is thought to be dammed-up underwater prey capture, a common behavior in the wild for which captivity offers no opportunity (Gewalt; Lyall-Watson in Grzimek, 1972, vol. 12, p. 99). The examples bear out that capture and killing movements may be performed when there has been little opportunity to exercise such behavior. These movements have their own sources of motivation, which, when not used adequately, 'go off' in vacuo. Such an explanation is supported by the examples mentioned being all derived from captive animals that are in some way deprived of stimulation normally provided by their natural prey.

It remains a puzzling problem why vacuum activities of the sort mentioned have not been found so far in invertebrates.

II. Prey Storing

Various birds and mammals tide over periods of food shortage by collecting and storing food when it is abundant. Storage habits are found in members of the crow family, in various tits and nuthatches (*Sitta europaea*), in shrikes (Lack, 1954, p. 139), in various owl species (summarized Räber, 1949; Scherzinger, 1970; Mueller, 1974), falcons (Mueller, 1974; Mebs, 1956), and red fox (*Vulpes vulpes*, Kruuk, 1964, Tinbergen, 1965); spotted hyenas, leopards (*Panthera pardus,* Kruuk, 1972 b: 119, 273), mountain lions (*Profelis concolor,* Hornocker, 1970), tigers (*Neofelis tigris*), lions (Schaller, 1967: 301, 1972: 271 seq.), and in bears (*Ursus arctos,* Mysterud, 1973). Only in animals that hide or carry off surplus food to a cache can one talk of storage. All the carnivores mentioned, with the exception of fox, tend to hide a carcass of a prey animal if they cannot consume all of it undisturbed, but they do not store several prey at a time. It will be examined to what extent storage of food, and the activities leading to it, are influenced by satiety.

The hungrier they are the more captive ravens (*Corvus corax*) hide pieces of meat or other prey items (Gwinner, 1965). Captive loggerhead shrikes (*Lanius ludovicianus*) impale them on thorns, a habit characteristic of the shrike family. But they do this increasingly the more sated they become, irrespective of the size of the food morsels ingested; after the tendency to impale reaches its peak it decreases again so that the behavior then resembles that of Gwinner's ravens (Fig. 7 a, b). The behavior of the ravens could also be interpreted to mean that they become inclined to hide more food as the day advances, and not that hiding increases with hunger. By carefully noting the hiding activity of one individual at the same time each day Gwinner showed that hunger was the crucial factor; the raven stored strikingly less food when pre-fed. Perhaps storing food in the wild is of advantage because food is distributed unevenly. Before more details are known about food storage by raven in the wild their different organization in this respect as compared to the shrike cannot be understood fully.

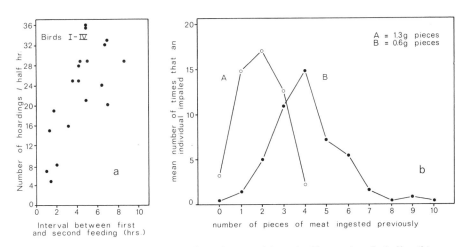

Fig. 7 a and b. Food-storing behavior of raven (a) and of loggerhead shrike (b) as a function of hunger as indicated by the length of deprivation (a) or degree of satiation in the course of feeding on morsels of known size (b). (Redrawn from Gwinner, 1965; Wemmer, 1969)

23

In the American kestrel caching food was found by Mueller (1974) to increase with length of food deprivation. He believes that this is caused primarily by 'previous frustration of anticipated feedings' and secondarily by hunger. The frustration hypothesis, however, is difficult to maintain in view of the fact that feeding intervals were 'capriciously intermixed' so that the birds *never* knew when a meal was due to come and thus could not have been frustrated (unless frustration is assumed to be permanent). Likewise the frustration hypothesis could not explain why caching continues on a regular feeding schedule. Moreover, hunger alone is sufficient to explain the observed increase of caching with deprivation. Mueller (1974) observed further that retrieving cached food under very restricted laboratory conditions remains constant over a considerable range of deprivation but depends on the time of day. This is hard to understand since one would assume retrieving to increase with hunger.

The performance of retrieving was remarkably good; the kestrels made few spatial and temporal errors, the latter consisting of an attempt to retrieve after not having cached. These errors did not bear a clear relation to the length of deprivation and, hence, hunger.

Redbacked shrikes (*Lanius collurio*) impale prey singly in their larders; reports are at variance whether storage occurs only during an abundance of food and during fine weather so that they can exploit their larder during wet weather (Schreurs, 1936; Münster, 1958). How this ties in with Wemmer's (1969) findings for the loggerhead shrike remains to be seen. In tits storage of seeds seem to depend entirely upon their abundance, regardless of what other conditions prevail (Löhrl, 1955).

Ravens (Gwinner, 1965) and pigmy owls (*Glaucidium passerinum*) (Scherzinger, 1970) store prodigious amounts of prey at times when they feed their young, especially after hatching. Thus storing activity appears to be activated by the motivation underlying parental care and, in the raven at least, by hunger.

Towards and during the winter, pigmy owls deposit their prey in larders in tree holes. Contrary to earlier authors, Scherzinger (1970) doubts the value of stored prey for times of prey scarcity because he found that the owls still continued to store birds in their larders in January and only began to utilize them from February onwards; moreover, he found that the owls had already depleted their larders long before the heavy snow cover disappeared in April. Both objections against the survival value of larders, however, rest on the premise that the storage mechanism would be as precise as possible, which is clearly unwarranted: Selection may be based on even minute advantages. As in ravens, larders may be of significance for the successful raising of offspring. This is even more pronounced in the thick-billed nutcracker (*Nucifraga caryocatactes*) which stores large amounts of hazelnuts in autum on which it feeds its young in the following spring when snow still covers the ground. A good crop in one year even tends to increase clutch size the following spring so that caching nuts has a double advantage (Swanberg in Lack, 1954: 30).

There is evidence that the motivation leading to storage of food differs from that underlying feeding in still other respects than described so far, i.e. in handling. Great grey shrikes (*L. excubitor*), store prey by wedging them into crevices. When about to store, shrikes tear prey animals into much smaller pieces than if feeding immediately; perhaps the ultimate reason is that morsels are edible even when frozen later on (Ullrich, 1971). Furthermore, in autumn moles (*Talpa europaea*) paralyze large numbers of earthworms by biting them in the head region and then store them in a larder, obviously in order to utilize them in winter (Skoczén, 1970; see also Yudin, 1972; *Asioscalops altaica*); but the picture is somewhat confusing as also uninjured earthworms are found, either together with damaged ones or by themselves. The injuries may not be the work of the mole (Godfrey and Crowcroft, 1960, p. 37). Also polecats (*Putorius putorius*) store frogs in larders, of more than a hundred in number which have been variously bitten and rolled about with the forepaws so that they can no longer move. It is not known to what extent polecats utilize their live stock of prey (Herter in Grzimek, 1972, p. 51, Vol. XII).

In raiding colonies of blackheaded gulls (*Larus ridibundus*) red fox kill exceedingly variable numbers of gulls on dark nights. The number of gulls cached on any particular occasion, on average 2.6 per raid, does not bear any obvious relation to the number of gulls killed, varying between 8 and 130 individuals (Kruuk, 1964). This seems to indicate that the tendency to cache prey not only fluctuates independently of hunger but also independently of previous kill performance.

In Conclusion. Storing prey may be influenced by hunger, yet the precise relations vary among predator species. Caching prey may vary both independently of hunger and number of prey killed before. In several predators handling of prey prior to storing differs from handling prior to eating, thus demonstrating that the motivation underlying storing differs from that underlying normal feeding responses. Up to the point of capture, causal factors are probably shared.

III. Providing Food for Dependent Family Members

Self-feeding and parental feeding are identical except for the end-point of the two behaviors. From a careful study of parental feeding in the oyster-catcher Norton-Griffiths (1969) concluded that both systems share common motivational factors, 'and that parental feeding cannot occur unless there is motivation for self-feeding' (p. 82). Unfortunately, there is no ground for testing the latter hypothesis since one cannot think of a bird capable of feeding young but unable to fend for itself. Norton-Griffiths noted that all oystercatchers, non-breeders and breeders alike, start feeding in the mud flats at the beginning of the low tide, generally cease to feed around the period of dead low water, and resume feeding again once the tide starts to flow and cover the feeding grounds. Parental feeding of young occurs only during the active phases of the feeding cycle. However, fluctuation in parallel to behavior patterns in the course of the day is far too rough a criterion to draw the conclusion mentioned; as a rule, activities of birds show basically the same bimodal ('Bigeminus' Aschoff) pattern of activity, without necessarily sharing the same motivational source. Instead, the observed daily pattern of feeding activity might be due to a corresponding vulnerability of prey animals (mussels, mud flat annelids, other terrestrial invertebrates), a hypothesis not yet sufficiently put to test.

Many birds feed their young with different food than they consume themselves, thus indicating motivational differences between both sorts of feeding. Many Middle-American birds (mainly Tanagridae) feed on fruit but provide their young with insects, presumably because of demands imposed upon nestlings to grow rapidly (Morton, 1973). Apart from such gross differences between self-feeding and parental feeding there are subtler ones. Royama (1970 a) has found that great tits (*Parus major*) provide the older broods with larger caterpillars than the ones they consume themselves (see also Verner, 1965: *Telmatodytes palustris*; Root, 1967: *Polioptila caerulea*; Kniprath, 1969: *Alcedo atthis*). Moreover, pied flycatchers and shama thrushes (*Copsychus malabaricus*) feed their freshly hatched young smaller-sized prey than they ingest themselves (see also Palmer in Siegfried, 1972). While cattle egrets (*Bubulcus ibis)* feed young nestlings with prey animals of roughly the same size as they consume themselves, they provide the nestlings with fish in a more digested state than later (Siegfried, 1972). While common tern (*Sterna hirundo*) males provide their mate during courtship feeding with larger prey, shrimps and fullgrown silversides (*Atherina* sp.), they feed their young at hatching the much smaller silversides, fry, and cunners (*Tautogolabrus adspersus*). The males that performed well at both periods were those that switched most successfully from one group of food items to the other (Nisbet, 1973).

By courtship feeding the female may receive substantial amounts of food. In the pied flycatcher it is half the feedings per hour, that a nestling (of the weight of the female) receives. Unfortunately, it remains unknown whether the food given equals the nestling food in terms of nutritive value (Curio, 1959 a). However, Löhrl (1974) made the revealing observation that the male coaltit (*Parus ater*) provides its mate with live insects as soon as these become available and even before it consumes them itself. If only seeds, even favorite ones, or dried-insect food are available courtship feeding does not awake. Before the eggs are laid the male takes the initiative while later the female solicits food by begging. Again the difference between two modes of feeding suggests differences in the underlying causation.

Apart from prey selection, differences between self-feeding and approvisioning offspring are shown in handling prey. In many birds of prey and owls the quarry has to be torn into suitable morsels which young nestlings are capable of swallowing. Curiously, whitethroats (*Sylvia communis*) decapitate flies before they feed them to their nestlings (Sauer, 1954). The orientation of food that is difficult to swallow is likewise adapted to an easy transference of prey. For instance, kingfishers orient a fish they are going to ingest in such a manner that the head is swallowed first whereas a fish for a nestling is offered head first (Kniprath, 1969).

Thus self-feeding and approvisioning of family members may differ in several respects, and capture and killing techniques inasmuch as necessitated by this difference. These facts support the view that self-feeding on one hand and parental-plus-courtship feeding on the other may share common causal factors only to the extent that the behavioral performance of obtaining the food is identical.

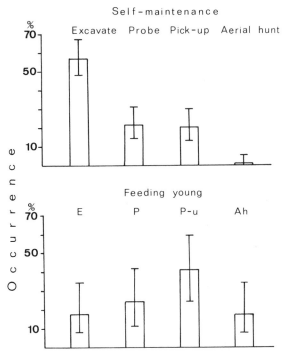

Fig. 8. The relative frequencies and their 5% confidence limits of foraging methods of self-feeding and feeding young in the Syrian woodpecker. (From Winkler, 1973)

This view receives still more support from an analysis of feeding methods. Syrian woodpeckers employ their diverse tactics for searching and picking up insect prey for their brood more evenly than when caring for themselves (Fig. 8); the diversity of their hunting methods as expressed by Shannon's H [1], increases, though not significantly, from 0.42 IE (1.5 bit) to 0.57 IE (1.9 bit). At the same time, the exploitation of different vegetation strata by the parent birds becomes significantly more diverse (Winkler, 1973, Winkler pers. comm.). Similarly, parent blue-gray gnatcatchers augment their feeding diversity by extending their foraging trips into the ground vegetation and other sites normally not exploited and by collecting more items as they become available (Table 1; Root, 1967).

---

[1] $H = -\sum_{i} p_i \ln p_i$, where p_i is the proportion of all items represented by category i.

Table 1. Foraging diversity of the blue-gray gnatcatcher as assessed by Shannon's *H*. The values of *H* are based on samples of 125 consecutive capture maneuvers in each category. Substrates where prey was obtained are classified as foliage, bark, herbs, and air; the capture maneuvers as gleans, hovers, hawks, and tumbles. (From Root, 1967)

	Diversity of substrates		Diversity of maneuvers	
	Self-maintenance	Feeding Young	Self-maintenance	Feeding Young
May 10 – June 16	0.542	1.006	0.804	1.133
June 24 – July 27	0.729	1.142	0.923	1.214

In mammals, another vertebrate group with elaborate parental care, both systems of feeding also appear to be different as judged from a number of facts (Mech, 1970, p. 143 seq., Kühme, 1965). For example, canids and among the cats only the cheetah (*Acinonyx jubatus*) (Schaller, 1972: 308), feed their cubs by regurgitation of meat, the reverse of ingestion. Second, domestic cats (Leyhausen, 1973, p. 61 seq.) and cheetah carry live mice or especially small, live gazelle to their litter. In their presence the mother releases the victim in order to elicit a chase and kill from the cubs. If these fail the mother may repeat the whole cycle until the cubs succeed in capturing the escaping prey (Schaller, 1972, p. 309).

Finally, there is sound evidence that the approvisioning of food for the offspring is under the control from stimuli emanating from the young rather than under the control of hunger from the parents. First, parent birds of many species augment their feeding frequency with increasing brood size though not in proportion to it (review Lack, 1954; 1966). Second, the begging behavior of the young is crucial for this adjustment as demonstrated experimentally by v. Haartman (1953) in the pied flycatcher. By reducing a brood of seven young to two young and then by continually replacing these two once they had been satiated with two hungry young, he showed that the parents brought food at the same rate as if there had been seven nestlings in the brood. Upon placing six hungry young in auditory but not visual contact with a nest containing only one young, the parents brought food at a high rate to the latter even though it was soon sated. Similarly, the begging behavior of hungry wolf pups makes the adults rise instantaneously, even around midday when they are dosing (Mech, 1970, p. 145).

C. The Problem of Specific Hungers

In order to maintain nutritional homeostasis an animal needs to do more than just become satiated. It has to have a choice among food items, unless it is monophagous, which is an exception rather than the rule. Domestic chickens, laboratory rats, and monkeys strive to select a balanced diet with regard to vitally important salts (Richter, 1943 a, b; Hughes and Wood-Gush, 1971), vitamins (Rodgers, 1967; Rozin and Kalat, 1971), or proteins (Peregoy et. al., 1972). Similarly, a number of insects show a clear choice for protein-rich food over carbohydrate-rich food when needed for proper egg development (Dethier, 1967). If a rat eats a food deficient in a given compound and is then offered a choice between a deficient diet and a diet enriched with the missing compound it rapidly compensates for the deficit due to a hunger specific for that compound, e.g. calcium, sodium, or water; or due to an appetite for an arbitrary novel diet, containing, e.g. thiamine (review Rozin and Kalat, 1971; Hughes and Wood-Gush, 1971). The question then arises if a predator is a "particulate feeder" as opposed to the rat as an "amorphous feeder", as regards powdered food, and can develop a specific hunger for a certain prey type in preference to others. Such a preference, however, need not involve a nutritional deficit but may concern a behavioral imbalance: Diversity of prey demands a diversity of hunting (see Chap. 5 B. I., II.) or handling tactics, so that the problem is not only one of specific hungers in the classical sense; it is also one of performing certain behaviors or perceiving certain cues in preference to others. Despite this being so the appetence for a new hunting behavior, for example, may merely help to achieve the ultimate goal of varying the diet in terms of nutrition.

I. Switching of Prey

After having fed on a restricted diet, predators tend to switch to prey hitherto neglected. After *Pterophyllum eimecki* had ceased to eat *Enchytraeus* worms, "central prey excitation" (p. 6) rises immediately upon seeing cyclopids when these are offered as new food although the new level does not reach the original one when feeding on worms began. Moreover, it falls to zero much more quickly (Fig. 9). The immediate rise of central prey excitation is striking since the size of cyclopids stays far below that of worms (v. Holst, 1948). Both facts when taken together indicate that part of the hunger of *Pterophyllum* is a general one, and part of it is specific for certain kinds of prey.

Everybody intimately acquainted with the "food-obtaining" behavior of captive insectivorous birds knows how avidly insects are taken after the birds have been fed exclusively on some mixture of ground insects, biscuits, and other ingredients. Similarly, as in the fish mentioned, captive insect-eaters, 'become bored' with one type of insect, as compared to another, rarer type (Swynnerton, 1919). Insect species differ by many visual cues. But one cue alone may suffice to bring about switching of prey. American kestrels which had eaten nine laboratory mice of one color in succession, attacked the tenth mouse more quickly if it was of a different color than those earlier ones (Mueller, 1972, 1975). It is difficult to escape the conclusion that the aberrant color was taken by the hawk to mean a new and therefore nutritionally different food. Handling or taste may be another such cue. Captive chaffinches (*Fringilla coelebs*) cannot survive on one type of seed alone. Kear (1962) fed groups of birds only three different types of seed for twelve days before and after twelve days with a choice of seeds. With rape and hemp, five out of six birds took less of the seed in the postrestriction tests; with canary seed there was no apparent rejection later. Since the various types of seed necessitated different techniques of dehusking, the need for keeping up a variety of seeds may not have been determined by metabolic needs alone.

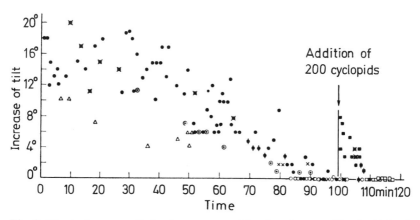

Fig. 9. 'Central prey excitation' as measured by the change of body tilt in the dorsal light reaction with light coming from the side as a function of prey species in *Pterophyllum eimecki*. *Left part of figure:* free feeding on *Enchytraeus* worms, in the *right part* on cyclopids. With the exception of o, prey was at least fixated and generally also eaten (●, ■). (From v. Holst, 1948)

Switching of prey may involve quite complex differences between prey animals. A clan of spotted hyenas hunts wildebeest for many days and suddenly switches to hunting Burchell's zebra (*Equus burchelli*); after that it will again hunt wildebeest and so on. These two ungulates form the mainstay of the prey. The temporal variation of prey preference is not just a chance result of encountering one or the other prey but the hunters are obviously intent on a particular prey species before they set out on a hunt. Hyenas intent on hunting wildebeest or Thomson's gazelle (*Gazella thomsoni*) go in a pack of, on average, 1.4 and 1.2 animals, respectively. By contrast, when intent on hunting zebra, packs number up to 27, with an average pack size of 10.8 animals (Fig. 10; Kruuk, 1972 b, p. 203). Moreover, when assembling for a zebra hunt, clan members go through quite a different and more excited ritual of 'greeting' than at other times. When leaving their den site they may pass hundreds of wildebeest or other would-be victims without paying the slightest attention to them until they encounter a herd of zebra (see also Schaller, 1972, p. 334 for wild dog); then they immediately start a chase. When a zebra is hunted down, pack size does not increase as it

30

does with wildebeest. Since the number of hyenas finally eating on a carcass is about the same with wildebeest and zebra, and since zebra weighs much more than wildebeest the reward per hyena after successfully hunting for zebra is correspondingly larger but needs a special social effort (Kruuk, 1972 b, p. 120 seq.). Whilst the switch from wildebeest to zebra might be explained by the hyenas having become hungrier from exploiting the less rewarding wildebeest, the reverse cannot possibly be explained this way; there is certainly no predator that prefers a light meal if it can have a big one. Accordingly, the switch from one to the other prey may be a matter of different stimulation or of hunting behavior; it is hardly due to a metabolic deficit. This view is forcefully supported by the existence of inter-clan differences in prey preferences which cannot be explained by differential prey availability in the respective clan ranges.

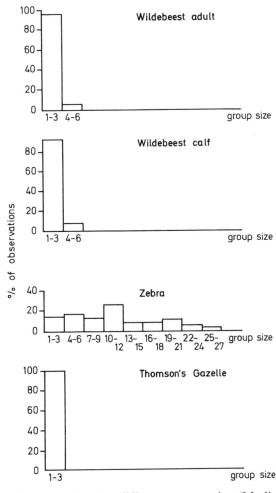

Fig. 10. Numbers of spotted hyenas in groups hunting different prey species. (Modified from Kruuk, 1972 b)

Wolves hunt in packs, remaining together over long periods of time, and do not seem to adjust pack size to the demands of a hunt but rather chase a prey animal which happens to be encountered en route, though they usually prefer prey that is easiest to capture. Though this may be the rule, switching in a pack from moose to caribou and back to moose over several series of respective kills (Mech, 1970) reminds one of "switching" in the hyenas.

A prey-specific appetence might also be expected in those predators which employ very different tactics of catching distinctly different prey. The staple food of the pigmy owl consists of small birds and mice which are hunted and killed differentially. A mouse is killed invariably with bites into the snout and the neck, a songbird by mere grasping with the feet, with still other differences in handling of the prey (Fig. 59, p. 172). Scherzinger (1970) believes that the motivation in both cases differs. (1) The appetitive behavior differs in that a pigmy owl intent on hunting mice perches on top of a tree scanning the ground below, whereas when sitting in ambush for birds it hides in tight bunches of conifers. (2) A pigmy owl which has been fed to satiation with live mice exhibits striking appetitive behavior for bird hunting and may even go through the act of killing: Though live mice are running about on the ground it dashes into the foliage, after an initial wing rattle and tail flicks as in a real hunt, grabs twigs, and mandibulates picked-up leaves. In which sequence pigmy owls in the wild prey on mice and birds is not known. A more thorough study of how palatability, handling of prey, hunting method, and metabolic needs contribute to regulating the two types of hunt would be rewarding. A similar qualification applies to piracy in birds which might be due to a motivation in its own right rather than to hunger alone (p. 161).

The nature of switching has been most fully explored for drinking behavior of laboratory rats (Morrison, 1974). The animal switches from one solution to another irrespective of which of the two solutions is the more palatable. Postingestional effects can be ruled out as a possible explanation. Results are compatible with the assumption that palatability of a solution declines as the animal drinks until it has fallen below the palatability level of a less preferred solution offered simultaneously. This occurs before satiation. Accordingly, the less palatable solution will be consumed until the animal switches back to the previous one and so forth. The mechanism might enable the animal to self-select an adequate diet from a number of choices, since it would allow for the alternation among foods of different palatabilities in the absence of specific deprivations. It remains to be seen whether this explanation could also account for the switching of prey by predators.

II. The Prey-density Predation Curve

Vertebrate predation generally increases as a function of prey-density in a characteristic way, described by Holling (1965) as the type-3 functional response curve (see Fig. 22, p. 60). While the lower part of the curve will be discussed later (see Chap. 2. C. IV. 1.), the levelling off to a plateau demands explanation now. As can be seen from Fig. 22 the total amount of food eaten remains constant over all prey-densities. This means that as the density of favored prey rises, the amount of the less favored alternate food eaten drops. What is more important for the present discussion is the fact that even at the highest prey densities the animal is taking at least some alternate food. This has been taken to represent the motivation of a predator to maintain a nutritionally balanced diet (L. Tinbergen, 1960; Kear, 1962; Soane and Clarke, 1973). Holling (1965) offered further evidence that the plateau of the curve is strikingly influenced by the palatability of the alternate food present; for instance, with the more palatable sunflower seeds the plateau will be reached much earlier than with dog biscuit. As a possible causal mechanism underlying homeostasis the predator is thought to have different thresholds of response for different food types. Briefly the hypothesis implies that the less favored prey will be accepted if the favored type is not encountered, if hunger has risen to reach the higher threshold of response. Thus if the thresholds of the two prey are not too different from one another, and the second prey is not too rare, the predator will automatically maintain a varied diet. Holling's hypothesis has been criticized by Krebs (1973 a) who pointed out that great tits exhibit a sigmoid functional response curve with regard to certain prey fed to their nestlings; so their own hunger level should not influence the food they choose. Moreover, the assumptions on the relative abundance of the favored and the alternate prey do not seem to hold true. As will be seen later, however, a second explanation of why great tits happen to have a varied diet can be offered (see Chap. 2. C. IV. 1.). Hence, the evidence for the existence of specific hungers is equivocal if based on the prey-density predation curve.

III. Swamping the Appetite of Predators

Certain protective devices of prey animals obviously take advantage of over-eating in predators. "Arithmetic mimicry" was suggested by van Someren and Jackson (1959) and by Glass (1959)[2] for the similarity between edible, yet conspicuous prey species, a type of mimicry fundamentally different from the ones surveyed so far (Wickler, 1968). The former authors found that members of the Pieridae and the Lycaenidae resemble each other and fly together. They assumed that the combined mortality rate is shared as a common burden of attack by predators, so that there is "safety in numbers". The selective force favoring resemblance would lie in the fact that the proportion of any one prey species destroyed would be less than in the absence of "arithmetic mimicry". For the model to work one has to as-

---

[2] The idea actually goes back to Wallace (1889, p. 245) who in this way explained the resemblance of the female *Leptalis melite* L. to one of the common Brazilian Pieridae, both of which species are edible and do not share a Batesian model species.

sume that the attack of predators is limited by satiation. There are two possibilities for prey species to swamp predators by sheer numbers, namely, either by swamping a specific hunger or a general hunger which is the sum total of all specific hungers.

As Holling (1965) has rightly pointed out, however, learning by vertebrate predators may promote massive diversity among prey species (Chap. 2. C. IV. 1., 2.). Such learning would oppose the effects of satiation invoked by van Someren and Jackson (1959) under certain conditions. As long as the combined population densities in a similar-looking swarm of prey species (taken by predators to be one species) stay below the plateau of the prey-density predation curve (Fig. 22), arithmetic mimicry would be disadvantageous and should give way to diversity (either within or between species). Only after combined prey-densities had reached that plateau would resemblance between prey species be favored by over-eating of predators. Holling (1965) doubts whether in nature, prey density will ever surpass the threshold for a selective advantage to accrue to arithmetic mimicry. Clearly, the idea ought to be critically tested in nature. It receives indirect support by the occurrence of impressive synchronization of reproduction in certain prey insects and ungulates.

Synchronization is thought to be adapted first, to an optimal food situation, and second to the mentioned swamping of predators. It must be stressed, however, that it is far from clear whether the predators merely fill themselves up to repletion, or whether a specific appetite gets satisfied through a massed food supply. For example, the periodic cicada (*Magicicada* sp.) emerge synchronously in such tremendous numbers that insectivorous birds are soon satiated (Lloyd and Dybas, 1966) though by their loud 'drumming' the cicada may protect themselves in still another way (Simmons *et al.*, 1971). Given huge populations, African migratory locusts (*Locusta migratoria*) derive similar protection from sheer numbers (refs. in Lloyd and Dybas, 1966). Similarly, the calving season of Thomson's gazelle and of wildebeest is synchronized to an extent that their predators are replete with fawns and calves which, once discovered, make easy prey. Gorging makes predators lazy and they rest for long times from which the as yet undiscovered segment of the population is thought to benefit (H. and J. van Lawick-Goodall, 1970, p. 117, Kruuk, 1972 b, p. 167; see also Pritchard, 1969: *Lepidochelys kempii*). [Wide spacing in time by prey species, on the other hand, could also be of advantage since predator populations cannot specialize upon the prey or cannot build up large populations (see Beattie *et al.*, 1973)].

D. Daily and Annual Rhythms in Predator-Prey Interactions

Predators are known to synchronize their predatory activity with the main activity of their prey. This in turn opens up the possibility for the prey to develop defenses based upon a daily rhythmicity.

34

I. Daily Rhythm of Predation

An example of synchronization of the predator's activity with that of the prey is provided by the hunting behavior of the pigmy owl in Finland (Fig. 11). The bank vole (*Clethrionomys glareolus*), which, apart from birds,

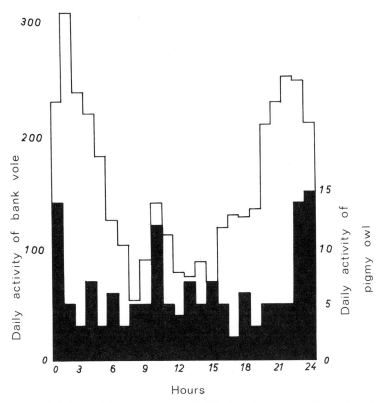

Fig. 11. Comparison of daily activity of pigmy owl (*black columns:* number of nest visits) in June 1969 and that of a female bank vole kept in natural conditions (*white columns*), 21 – 27 July and 1 – 6 August 1969. Sunrise and sunset were approx. 1.4 and 23 hours for the owl observations, and approx. 3 and 21 hours for the vole recording. (From Mikkola, 1970)

forms over half of its diet (Jansson, 1964; Mikkola, 1970) exhibits a bimodal ('bigeminus' of Aschoff) pattern of endogenous (Erkinaro, 1972) activity which coincides with that of approvisioning visits to the nest of a pigmy owl. Similarly, the nocturnal acitivity of a cat was recorded in captivity under a natural light-dark regime and an irregular provision of food during the day. The duration of nocturnal acitivity almost exactly matches that of one of its staple prey animals in the wild (*Apodemus sylvaticus*; Schuh *et al.*, 1971). It must be stressed though, that relations such as these merely suggest

35

that the activity of the prey is the major causal factor involved that has brought about the daily pattern of predation.

In two other cases the daily pattern of prey selection appears to be triggered by the pattern of prey availability and vulnerability. In a unique way, spotted hyenas switch from scavenging from lions' kills in the day to hunting successfully on their own in the night. Kruuk (1972 b, p. 110 seq.) believes that this is due to the hyenas observing events around a fresh kill, e.g. the descending of vultures to participate from it etc., which they obviously cannot make use of in the night. Another factor contributing to this pattern of predation may be the greater vulnerability of the hoofed prey animals in the dark. This feeding pattern is by no means rigid, as hyenas around human settlements can take largely to scavenging from offal. Moreover, the flag cabrilla, a marine piscivore, feeds regularly during the whole 24 h. It preys on fish during the day and on crustaceans in the night when these emerge in large numbers in the sandy expanses offshore (Hobson, 1968). Similarly, many 'nocturnal' feeders among marine fish will readily feed in the day if and when food becomes available (Starck and Davis, 1966). These observations suggest that the predatory pattern of daily activity is, at least in some cases, pretty much adjustable to the opportunities to feed. It is therefore not conceivable to think rigidly of activity patterns in terms of 'diurnal', 'nocturnal', or 'crepuscular', although this terminology still fulfils practical purposes. Apart from these facultative shifts of the main activity within the 24-h cycle the picture is complicated by species with many short 'bursts' of activity alternating with about equally long periods of rest. Many mammals, especially rodents and insectivores, have been found to be active in 'bursts' with a period length of 2.5 or 3 h while a 24-h activity pattern is still recognizable in that these bursts are more frequent, more intense, and of longer duration in the night than during the day (review Aschoff, 1962; see also Erkinaro, 1973; Zbinden, 1973: *Serrasalmus nattereri*). A discussion of details of the daily activity pattern is outside the scope of this book. However, a few remarks appear appropriate at this point. Field observations have suggested that a number of animals are active during the whole 24 h. In this group belong such diverse creatures as certain geckos, vipers (*Vipera xanthina palaestinae*) (Gruber 1971: 165 seq., and Petzold 1971, p. 451 seq., both in Grzimek 1971, vol. VI), and mountain lions (Hornocker, 1970), i.e. species which have been seen to hunt at any time of the day. Yet the crudeness of the observations does not rule out the existence of a pattern of bursts which are superimposed upon a 24 h cycle as sketched above. Despite this uncertainty, the question arises as to whether individuals of such species alternate with each other in their periods of activity and rest so that activity cycles, both individual and when viewed together, result in a composite picture of activity all around the clock. A possible mechanism could be that individuals exclude others from their own segment of the day so that there is 'territoriality' in terms of time (see below the domestic cat). As a matter of fact, European moles have been found active, by means of radioactive marking and a Geiger counter, in bursts of 4.5 h followed by periods of rest of (at least) 3.5 h. There was no synchrony in the activity of different individuals but this lack of synchrony is presumably not due to an interaction between individuals: the same mole was found active at different times on successive days (Godfrey, 1955, Godfrey and Crowcroft, 1960, p. 58 seq.). Another question that arises is whether species with short-term fluctuations of their activity are more prone to facultative shifts of their main peak of feeding activity as a result of altered food abundance.

Likewise, details of temporal predatory organization are but poorly understood. Kavanau and Ramos (1972) found in three nocturnal carnivores (*Potos flavus, Genetta genetta, Bassariscus astutus*) that the onset of daily activity coincided with dusk but that it terminated quite variably and long before dawn. This pattern is believed to maximize predatory success and avoidance of predation, but it is far from clear in which way the recorded pattern properties might contribute to these ends. The same pattern has also been observed in nocturnal birds, and has been explained by Aschoff and Wever (1962) as an inevitable outcome of inherent properties of the clock mechanism underlying circadian rhythms. Moreover, neither the 24-h activity patterns in a number of animals (e.g. Holling, 1966; Beukema, 1968; Lack, 1954, p. 136 seq.) have yet been understood functionally, nor is there any set of 'grand rules' describing the interplay of such rhythms with hunger (see Chap. 1. B. III.; Aschoff, 1962).

The daily rhythm of predatory activity, if any, can be adjusted to demands imposed by other predators, either of the same species or of others. Domestic cats ranging free use territories and trails of a common hunting area in space and time, together with several other individuals, but they avoid contact with the others. However, while such avoidance may take the form of a truly daytime-related exclusiveness in captivity (also Steiniger: *Ondatra zibethica*, pers. comm.), it is rather a matter of eye-sight in the wild (Leyhausen, 1973).

Among different species of predators the higher ranking one may impose a regimen of periodicity upon the biologically inferior. On Galapagos Islands where the Galapagos hawk (*Buteo galapagoensis*) is absent the short-eared owl (*Asio flammeus galapagoensis*) is active both day and night whereas on other islands it confines its activity entirely to the safe nighttime hours; if it dares to take wing in the presence of the hawk during the day it will be attacked immediately (de Vries, 1973; see also Meinertzhagen, 1959, pp. 135, 137). It appears entirely possible that interactions of this kind have contributed to the habitually crepuscular and nocturnal life of the Strigiformes; of 133 species some 80 hunt during the night and some others around dusk (Sparks and Soper, 1970, p. 154), whereas only very few members of some 280 Falconiformes (*Machaerhamphus alcinus, Falco rufigularis, F. subbuteo, Elanus scriptus*) are crepuscular or nocturnal, perhaps because of the corresponding life habits of their bat and rodent prey (Brown and Amadon, 1968). Along similar lines, Hailman (1964) speculated about the nocturnalism of the Galapagos swallow-tailed gull (*Creagrus furcatus*) as being a defense against piracy from the common frigatebirds (*Fregata* sp.).

II. Daily Activity Patterns of the Prey

In order to evade predation, prey animals may employ greater measures of defense during that time of the day when they are more endangered, or, they shift the phase of their daily activity away from that of their pursuer, a possibility which has apparently been made use of by marine schooling fish (Clupeidae, Pomadasiidae, Carangidae, Lutjanidae, Sciaenidae, Mullidae). As carefully documented by Hobson (1968, 1972) diurnally active species go to seek shelter for the night just before sunset while nocturnal species head from inshore resting places to offshore feeding grounds at the end of twilight; likewise prey fish exclude the morning twilight period from their longer-range movements, so that during each twilight period there is a 'quiet' or 'interim' period with little prey movement. It is scarcely mere coincidence that fish suffer most heavily from attack by the large piscivores (*Mycteroperca, Caranx, Elops, Nematistius*) during these 'quiet' twilight periods (see also Starck and Davis, 1966), presumably because these piscivores have to compromise between visibility necessary for attack and avoidance of untimely discovery by the prey. Excluding the evening and morning twilight periods from their diel pattern of activity, as evolved by many families in parallel, seems to protect them against crepuscular enemies (Hobson, 1968). Analogous examples from diverse prey groups have been enumerated by Remmert (1969). Such avoidance (see also Loop, 1972) becomes futile if groups of predators like owls and raptors take turns around the clock as mentioned; it will then be impossible for the prey species involved to evade one or the other.

Another means of defense is diel variation of alertness. In a preliminary overview Jouvet (1967, p. 335) stated: "We find a fairly strong indication that the hunting species (man, cat, dog) enjoy more deep sleep than the hunted (rabbits, ruminants), in our test the former average 20% of total sleep time in paradoxical sleep whereas the latter average only 5 – 10%. Further studies are needed." The coral-reef fish *Chaetodon melanotus* reduces swimming activity during twilight but becomes ready to dart off in long dashes upon the slightest disturbance. During the night its behavior becomes sluggish and the fish unresponsive to alerting stimuli (Fricke, 1973). The periods of both increased alertness and reduced swimming activity coincide with the peak of predatory activity of piscivores (see above).

Movement is a pervasive and powerful prey-attack eliciting stimulus (p. 89). Hence, both behavioral changes of *Chaetodon* appear to be anti-predator adaptations. Similarly, the response of common fresh-water fish (*Leucaspius delineatus*) to danger stimuli changes during the 24-h cycle (Fig. 12). While schooling increases in the day in response to pike, both free and behind glass (B, C), and to conspecific alarm substance (D), it remains unaffected or inconsistent in response to free tench (E). By contrast, in the night a flight response to all stimuli mentioned overrides any tendency to school as compared with nonstimulation; tench also have an effect (compare D and E), presumably because they damage and eat *Leucaspius* and thereby cause alarm substance to become released, though this may not be the only explanation (Rüppell and Gösswein, 1972).

Particular protective behaviors that are at work only during the daily period of imminent danger constitute a second line of defense. Upon mechanical disturbance a nocturnal holothurian discharges its toxic Cuvierian tubules only during the day when the animal is certainly most subject to actual attack by fish predators. There is conflicting evidence as to this diel pattern in the course of the year (Bakus, 1968). The fifth instar larvae of a neotropical hawkmoth (*Erinnyis ello*) display while walking a peculiar rocking behavior only during the day. During that time they are preyed upon by visually oriented predators. As preliminary tests suggest, rocking is

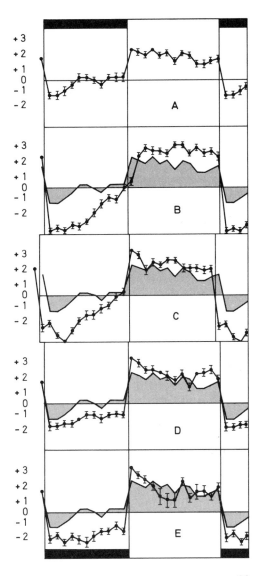

Fig. 12. Schooling (0 to +3) and flight behavior (0 to −2) in 350 *Leucaspius delineatus* to various danger stimuli as a function of time of day. Averages and standard deviation from 5 days each. *A* undisturbed; *B* in the presence of 3 free pike; *C* 3 pike in a transparent tank with in the test fish tank; *D* alarm substance; *E* in the presence of 3 free tench. Hatched: Reference curve *A* (From Rüppell and Gösswein, 1972)

39

bound to ambient illumination; even in the night, when the behavior is normally absent, switching on a light can artificially elicit it. Whether rocking is controlled solely by external stimuli remains to be investigated (Curio, 1965 a, and unpublished data). Rocking movements have been shown to be of definite survival value in stick insects when visually hunted for by marmosets (Robinson, 1966).

Protective countershading (de Ruiter, 1955) is achieved in the pony fish by employing bioluminescence in place of light coloration on its ventral side. As discovered by Hastings (1971) the fish switches off the ventral glow over night when it would tend to produce just the opposite effect of what it presumably has been designed for, i.e. to conceal the fish. It may be significant that quite diverse luminous animals, especially crustaceans and squid (*Loligo* sp.), also display ventral bioluminescence (Allister, 1967 quoted by Hastings, 1971).

An unresolved puzzle of daily rhythmicity is posed by the flight response of fresh water snails to body juice of crushed conspecifics. As reported by Snyder (1967) the response can be elicited both during day and night in species which flee by leaving the water whereas in others, which flee by burying into the sand, it can be released almost only during the day. Likewise enigmatic appears the fleeing response of the Florida apple snail (*Pomacea paludosa*) to odors of its main turtle predators (*Sternotherus minor, Chelydra serpentina*). The snail falls off from its substrate and buries rapidly into the sand when sensing turtle odor, yet, surprisingly, only during the day although hunting prevails in the night. Strangely, the burial alarm response to juice of conspecifics is not inhibited by darkness. Darkness affects the turtle response through absence of light rather than by photoperiod: if, in the presence of turtle odor, lights are switched on a few minutes after darkness, part of the snails still bury themselves. (Snyder and Snyder, 1971). The imperfection of this highly stimulus-specific response is hard to understand since it exactly matches the needs in other respects.

While these relations seem to be obligatory, there are others where only the actual presence of a predator causes the prey to be wary. The mere presence, if perceived, of a cat may interfere with deep sleep in rats (Gibson, 1970). Also, disturbances may have undesirable effects a considerable time later. After a raccoon had caused nightly panics in a breeding colony of *Larus delawarensis* the gulls panicked in subsequent nights although the interloper did not raid in again. Their behavior during the day remained entirely normal (Emlen *et al.*, 1966). Other prey species are able to resort to some more adaptive response. Pheasants switch from roosting on the ground to roosting in trees after ground predators have become a source of disturbance (Glutz *et al.*, 1971).

As a rule, animals become more shy in parts of their range where they have been most often pursued. Reports on an actual shift of the phase of daily rhythms due to regular hunts of predators are noticeably scarce. Young of the Nile crocodile (*Crocodilus niloticus*) which in Natal were repeatedly attacked by African fish eagles (*Haliaetus vocifer*) during the day become entirely nocturnal while others, which were kept in eagle-safe enclosures, did not pay any attention to the attacks and remained diurnal as usual (Guggisberg, 1972, p. 82). Furthermore, teal (*Anas crecca*) when wintering in the Camargue, fly to their feeding grounds after dusk, stay there overnight and return before dawn. At their diurnal resting places they are subject to heavy predation from herring gulls (*Larus argentatus*) and marsh harriers (*Circus aeruginosus*); accordingly, because of their constant vigilance and frequent antipredator flocking the teal could possibly not feed even if the food were there (Tamisier, 1970). On other wintering grounds with much less predation pressure as for instance in Aarestau Klingnau (Switzerland) feeding is spread all over the day with some predilection for dusk and dawn (Bauer and Glutz, 1968, p. 363). Furthermore, pine marten may be diurnal in remote forests while they are shy and strictly nocturnal near human habitations (Löhrl, 1972; see also Gundlach, 1968: Suidae; Remmert, 1969: Brachyura).

III. Annual Rhythm of Predation

Among rhythms with a period length of more than 24 h circannual rhythms have come to be known as governing reproduction (reviews Aschoff, 1955; Immelmann, 1967), migration and molt (recent review Gwinner, 1972). The food composition of temperate zone species exhibits also marked annual fluctuation (e. g. Lack, 1954; Pimlott *et al.*, 1969; Smeenk, 1972). However, in virtually no case is it known if this seasonality is governed by an endogenous circannual rhythm or simply by the seasonally fluctuating availability and vulnerability of prey animals. Circumstantial evidence in birds indicates that hunting intensity and prey selection may vary annually in spite of near-to-constant external conditions. Pigmy owls commence massive killings of songbirds in midwinter for storing (p. 24); at the same time vacuum prey captures by captive individuals abruptly increase (Scherzinger, 1970). Great bustards (*Otis tarda*) have been reported to accept mice only in the summer but not in the winter (Radu, 1969), a feeding pattern that may be adapted to the seasonal vulnerability of mice. Furthermore, in the fall great grey shrikes hunt more after insects than after common lizards (*Lacerta vivipara*) although the availability (and vulnerability?) of lizards remains the same all the time; while insect prey increases from 52.2% during the nestling time to 94% in the fall, the percentage of lizards in the diet drops from 26.7% to 0% (Grönlund *et al.*, 1970).

The only study conducted so far under conditions of constant photoperiod is that of Berthold and Berthold (1973) with hand-raised garden warblers. These birds exhibit a circannually changing food preference in that they strongly prefer mealworms to berries during their deposition of fat for migration. Thereafter the birds switch to a more balanced diet of both foods while finishing their winter molt. Berries alone cannot maintain the energy budget for any length of time. Facts as these would be further evidence in favor of the hypothesis of specific hungers in predators (see Chap. 1. C.).

Anti-predator behavior of the prey is known to fluctuate in the course of the year (review Curio, 1963) but one that specifically meets the demands imposed by an annually changing hunting behavior seems as yet unknown.

Chapter 2

Searching for Prey

Searching brings a predator's exteroceptors within the range of stimuli emanating from potential prey. The movements employed during searching for prey can rarely be distinguished from those shown in other search behavior so that only by the act of prey capture can the preceding behavior be labeled searching. It is most appropriately seen as hunger-dependent selective responsiveness to some stimuli as opposed to others (see p. 14).

A. Path of Searching and Scanning Movements

Animals in nature are distributed patchily with species restricted to a limited range of habitat units and with individuals aggregated or exhibiting "contagious distribution" (Southwood, 1966, p. 24 seq.). The dispersion of potential prey and its variations in density create a number of spatial problems for predators; they must locate prey concentrations efficiently in order to feed economically.

The most thorough study of searching behavior was conducted by J. Smith (1971) using the blackbird (*Turdus merula*) at Oxford. Analysis of precise records of the blackbirds' tracks while searching for food on a large meadow led to the formulation of "rules" describing the path before and after capture of a prey item. The well-known pattern of a blackbird searching on a meadow for invertebrate prey consists of roughly linear moves and of stops during which it performs obvious scanning movements with the head. Mean move length was 38 cm with an unexplained difference between the sexes. The turns after a stop were symmetrical about the zero direction provided by the heading of the previous move, with 67% of all turns lying within 43 degrees of this heading, and turns of opposite sign tending to alternate. It seems that move length and turn size were determined independently of each other. The main question remaining unanswered is: to what extent are moves and turns decided upon by some generalized tactics and by current environmental stimuli? Both moves and turns appear to be determined by no more than the two moves previously made.

It should be evident that a predator searches more efficiently the better it is at avoiding an area just covered. The ongoing movement path of the blackbird is clearly effective in avoiding the pitfall of random movement, though it is unclear if this form of path has been specifically designed to avoid waste of energy expenditure: Fully sated goldfish show strikingly similar distribution patterns of moves and turns and also a tendency to alternate right and left turns, when swimming in a large uniform tank (Kleerekoper *et al.*, 1970). However, as we shall see the sequential turning pattern in the blackbird can be modified as a function of prey distribution. For insects such as ichneumonid parasitoids avoidance of a host which has been detected by a competitor is of vital importance. Here it is only by odor trails that reliable avoidance of an area or a host just visited has been achieved. As Price (1970) has demonstrated females of four parasitoid species avoid an area not only presearched by themselves but also by females of another species or genus, thus ensuring that only one egg be placed in a host (see also F. Wilson, 1961).

It remains unknown in which way scanning movements of the eyes, readily amenable to investigation in humans (see Noton and Stark, 1971), contribute to the detection of prey. Domestic chicks lower their heads below the level of the top of their back prior to pecking, when tested for their ability to detect conspicuous or cryptic grains of rice both of which they are familiar with from a noncryptic situation. Whereas the first conspicuous grain was taken after merely 1.3 sec in the head-down posture, the first cryptic grain was picked up after 33.5 sec ($P = 0.0013$). The chicks also pecked more at stones when searching for the cryptic grains. It would seem then that cryptic grains were not seen when the chicks first lowered their heads and looked at the ground. As suggested by these crude measurements and as demonstrated by thorough experimentation (p. 64) some further, possibly central change would have to occur before chicks were able to detect the cryptic grains (Dawkins, 1971 a). It is likely that the longer period needed for detection of cryptic grains was used for more scanning movements with the eyes.

Whether the size of the perceptual field tends to vary as a function of hunger (Chap. 1. A. IV. 1.) or experience with prey (Chap. 2. C. I. 2.) remains to be investigated. Paloheimo (1971), in a stochastic theory of search, assumed that the perceptual field of a searching predator was circular, though this need not be so: in a mantis it is elongate (Chap. 1. A. IV. 1.). Moreover, the perceptual field cannot be deduced from anatomical considerations alone. A searching herring larva (*Clupea harengus*) notices planktonic prey organisms, mostly copepods, that happen to be in a "tube" extending laterally and above its swimming course (Fig. 13). Prey beneath the bot-

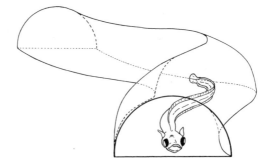

Fig. 13. 'Tube' searched by a herring larva while search swimming, a slow meandering. (From Rosenthal and Hempel, 1970, by courtesy of Oliver and Boyd and Otto Koeltz Antiquariat)

tom plane of this 'tube' will be ignored (Rosenthal and Hempel, 1970). European jays (*Garrulus glandarius*) searching for insects in the foliage, tend to detect resting caterpillars from below rather than from above or at eye level (de Ruiter, 1955). Yet this may also be a consequence of the larval resting positions.

The perceptual field may widen through rhythmic lateral displacements of head or body. Herring larvae adopt a "slow meandering" while search-swimming, with a large horizontal amplitude of only very little net progression, contrasting with two other ways of swimming; thereby the field of vision across the path of a searching larva increases by a factor of 1.5 (Rosenthal and Hempel, 1970). Similarly, coccinellid harvae (Fig. 14) while stop-

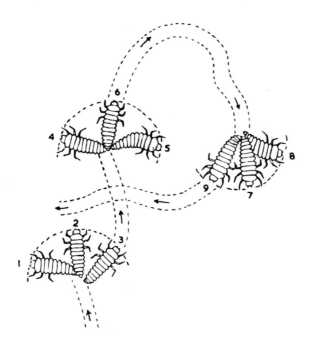

Fig. 14. Sideways-directed 'casting' movements showing 'arc of searching' behavior by a coccinellid larva. (From Banks, 1957)

ping in the search for aphids, perform alternating sideways "casting" (see also Chandler, 1969) movements with the forepart of the body raised, thereby seeking to physically contact prey within an arc of 160–220 degrees. Searching of syrphid and chrysopid larvae is similar (Bänsch, 1964).

Probabilitiy of capture success depends upon whether recognition of prey is made by contact or by perception at a distance, and secondly, upon the movements of predator and prey relative to each other. If the predator searches over a two-dimensional surface for fixed prey, as is generally the case in insect parasitism, probability of contact is given by the formula (Laing, 1938):

$$N = \frac{1}{G} [\mu \gamma (\sigma_1 + \sigma_2)] \tag{1}$$

where:

N = probable number of contacts per unit time
G = total area available for search
μ = the mean speed of movement of the predator
γ = the number of stationary prey
σ_1 = mean diameter of the prey
σ_2 = diameter of the predator

When the predator recognizes the prey at a distance, σ_2 becomes the diameter of the perceptual field (Holling, 1965, 1966). From the above considerations it should be clear that what a herring larva does when meandering, or a syrphid larva when "casting" sideways, is increasing σ_2. From Eq. (1) it follows that the probability of contact can be reduced by a prey by reducing its body size (σ_1). Evolutionarily, because of apparent limitations, this has been perhaps rarely achieved by prey species. Yet Roberts (1972) believes that the occurrence of minute fish species, developed by several families in parallel, is a means of escaping predation from piscivores where these abound as in the Amazon and the Congo basins. Though only of a temporary nature, crouching is a common anti-predator-device by which apparent body size shrinks. On the other hand, prey odor may enlarge apparent prey "size".

A slight modification of Eq. (1) allows an assessment of contacts with stationary prey within a three-dimensional volume which is assumed to be uniform in character (Salt, 1967).

If the predator searches for prey that are in continuous and random movement, a very common situation, Eq. (1) can be used, especially if the prey density is high, by assuming that the likelihood of a prey animal being at any given spot is the same as if the prey were stationary and randomly placed (Salt, 1967). Or, an equation may be used in which the speed of movement of both predator and prey are combined to give an average of their movement relative to one another (Holling, 1966):

$$N_E = [2\,(VR)\,t\,(r_A + r_B) + \pi\,(r_A + r_B)^2]n_B \qquad (2)$$

where:

N_E = number of encounters per unit time
r_A = radius of predator or its perceptual field
r_B = radius of prey
n_B = density of prey
VR = approximate mean velocity of the two $= (V_A^2 + V_B^2)^{1/2}$
t = time
V_A = velocity of predator
V_B = velocity of prey

This holds for predator and prey moving on a two-dimensional surface. A modified formula would describe movement in a volume. As in Eq. (1), an increase in body size of prey or of the perceptual field of the predator would facilitate its detection by the latter. It must be emphasized though that the discussion so far is an oversimplification since it does not take into account the common fact that prey animals also have a perceptual field within which they discover and subsequently avoid the approaching predator.

45

B. Area-concentrated Search

I. Short-term Area Concentration

After a searching predator has captured a prey it tends to stay in the vicinity of the capture. This feature of the hunting strategy has important consequences for the pattern of distribution of prey animals.

1. Living Scattered and Area-concentrated Search

The efficiency of search will be enhanced by any one or a combination of three changes of overt behavior: an increase in turning or of the width of search path, or, by slowing down forward movement. After a coccinellid larva has contacted and captured an aphid its path of search becomes more convoluted (Fig. 15). This change of path has been found in the larvae of other predatory insects as well, i.e. in other coccinellid larvae (Fleschner, 1950; Dixon, 1959; Bänsch, 1964), the neuropterans *Chrysopa californica* and *Conwentzia hageni* (Fleschner, 1950), syrphid larvae (Bänsch, 1964; Chandler, 1969; see also Sandness and McMurtry, 1972), and also holds for the search behavior of insect parasites (Laing, 1937, 1938). Since aphids, the staple prey of the larvae mentioned, usually live in clusters, the increase of turning after capture seems to be adaptive. Why clustering among aphids is so common may be functionally explained, in that it protects the colonies against certain insect parasitoids (p. 126).

Convolution of path also pertains to barnacle and serpulid polychaete larvae which investigate a suitable substrate by this means before settling for life in the best possible position (review Reese, 1964).

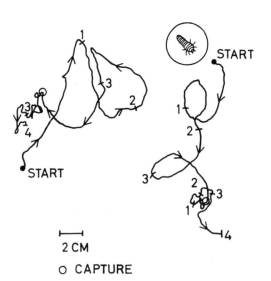

Fig. 15. Searching path of a hungry fourth instar coccinellid larva before and after capture ○ of a prey. Dashes across tracks correspond to 4 successive 15-sec intervals. (Redrawn by J. Smith from Banks, 1957)

46

Basically the same tactics are employed by herring larvae that adopt a slow meandering when entering a patch of their planktonic food. When food becomes scarce swimming becomes faster and straighter, thus leading the larvae into a new patch of food where they again stay longer by increasing their slow meandering (Rosenthal and Hempel, 1970). Similarly, the path traced out by some of the predatory larvae mentioned not only displays tighter turning; in addition the scanning "casting" movements with the forepart of the body (p. 44) tend to increase in frequency (e.g. Bänsch, 1964) while forward movement slows down (Chandler, 1969).

Camouflaged prey animals seem to avoid the neighborhood of conspecifics, i.e. they usually live 'scattered'. Living scattered confers a high selective advantage as compared to living clumped or clustered (see p. 51). The "Inter-prey distance" of spaced-out individuals is in many cases very much greater than the "Direct detection distance" from which a predator finally locates a prey individual. In order to account for this discrepancy Tinbergen et al. (1967), assumed that the predator exerts pressure upon prey ani-

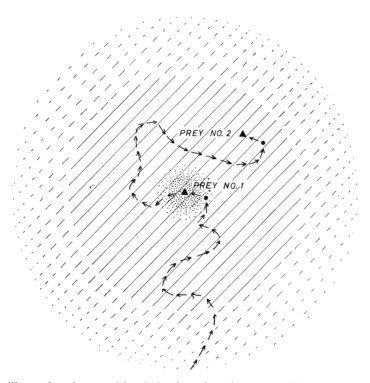

Fig. 16. Diagram illustrating the searching behavior of carrion crows (*Corvus corone*), and the consequent risk imposed on a prey item (*No. 2*) by the close proximity of another prey (*No. 1*). → track of crow. • place from where a prey is seen. Radius of stippled area represents the Direct Detection Distance, variations of which are indicated by the vagueness of this area's boundary. Area-restricted search upon finding prey *No. 1* increases the risk for any subsequent prey in the hatched area, the radius of which is the Effective Detection Distance. (From Tinbergen et al., 1967)

mals apart from its direct detection distance which is largely dictated by its perceptual field. They assumed that the crows which they had observed in the field when finding concealed eggs would alter their behavior into an "area-restricted search" near the spot of a find as described for the predatory insects above. The searched area surrounding a capture point was visualized to be circular and its radius called the "Effective detection distance" (Fig. 16; but see below), which is assumed invariably to exceed the direct detection distance. The idea of Tinbergen *et al.* also implies that the search becomes really 'area-restricted', otherwise no amount of scattering would avail.

As pointed out by J. Smith (1971) the term "area-restricted searching" is misleading as it seems to imply that the predator searches only the area surrounding a previous prey capture. In fact, however, crows, thrushes, and many other predators too, typically continue their search after moving off to other nearby areas. Instead, J. Smith proposed the term "area-concentrated searching". Significantly, as shown by Thomas (1974 and pers. comm.) in neatly controlled experiments, sticklebacks and bluebottles (*Calliphora erythrocephala*) not only exhibit this type of behavior after ingesting food but also "area-avoided search" upon rejecting a food item. In avoiding the area of rejection, sticklebacks swim faster and straighter, i.e. show just the reverse of area-concentrated search. Such behavior in nature would lower the likelihood of encountering unpalatable prey in rapid succession, provided it occurred in groups. The disparity of search and avoidance behavior reminds one of the swimming patterns of herring larvae in food situations of differing profitability. In conclusion, avoidance of an area takes place if the area has been depleted by previous exploitation (p. 43) or has proved to offer nothing but undesirable food.

In order to avoid ambiguity Croze (1970, p. 20) replaced effective detection distance by "search saturation". The risk imposed upon scattered prey by "area-concentrated searching" depends upon two parameters: (1) The radius R of the circular area around the find determining the "Search range". (2) The "giving-up distance" of the predator, that is the total length of path along which it is prepared to search without reward until it leaves the area. For a crow this means flying to a new place. The giving-up distance in turn determines the "Search saturation" which means the fraction of the search range actually covered (amended from Croze, 1970: 22):

$$\text{Search saturation} = \frac{\text{giving-up distance} \times f \times \text{direct detection distance}}{\pi R^2}$$

The numerator is the total area covered by the searching eyes with $1 < f < 2$; Croze (pers. comm.) believes that $f < 2$ because the crow is not scanning the area left and right of its path equally within the direct detection distance. All other things being equal, the smaller the radius R of the search range is, the more thorough the search for a standard giving-up distance. Conversely, a predator with R remaining constant can only increase its search saturation by lengthening its total giving-up distance, which in fact occurs (see below).

2. The Nature of the Path Changes

General. Amongst what little concise information on the actual searching behavior after prey capture is available Smith's (1971) painstaking analysis of prey-searching behavior in blackbirds is taken as an example. Figure 17 illustrates diagrammatically the tracks of a thrush before and after capture. Beeline distances which encompass an equal number of stopping points before and after capture are compared to each other. The results of this path

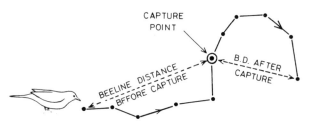

Fig. 17. Schematic presentation of method used to examine whether systematic path changes followed prey capture by blackbirds. Compare the 5 moves before and after capture. (From J. Smith, 1971)

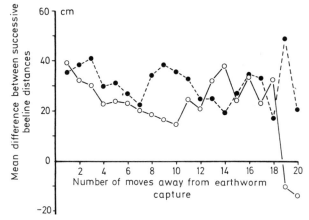

Fig. 18. Mean changes in beeline distance as blackbirds approached •, or moved away from the positions of earthworm capture ○. Fifty captures analyzed. Ordinate: change in beeline distance for each successive move away from the capture point, e.g. from capture point to first point before or after capture, from the first to the second point away from capture etc. (cf. Fig. 17). (From J. Smith, 1971)

analysis upon one measure (Fig. 18) show that beyond the first move after capture, the next eleven changes in beeline distance were smaller after capture than before capture. In other words, the blackbirds tended to remain near the capture point over the first twelve moves after capture thus confirming Tinbergen's original idea of area-concentrated search. Beyond this point, the behavior before and after capture did not show a consistent difference.

This result raised the question as to the nature of the changes that led to the shorter beeline distances after capture. Since there were neither changes in mean move length after capture, nor mean absolute angle turned, there must, by eliminating these two possibilities of beeline change, be changes in the sequential patterning of turns, i.e. the sequence of left and right turns probably changed after capture so that lefts and rights no longer tended to alternate (see Fig. 18).

The Influence of Prey Density and Distribution. So far the capture of a single prey item has been regarded as an isolated event. It will now be examined in which way the search method of the predator will be modified by a given prey density and prey distribution as to maximize exploitation. *Woodruffia metabolica,* a holotrichous ciliate, decreases its searching effort only slightly in response to a tenfold increase of the density of *Paramecium,* its only prey animal. Evenness of terrain is of greater importance; the searching rate is reduced to a third of the rate shown on a smooth clean surface if the bottom of the culture dish is covered with bacteria (Salt, 1967). In a classic experiment Laing (1937, 1938) used females of the ichneumonid parasitoid *Trichogramma evanescens* which searched for host (*Sitotroga cerealella*) eggs on a two-dimensional surface, with the total volume of confinement being kept constant. The distance between eggs (Inter-prey distance) was varied and the number of contacts comprising oviposition, examination or mere touch, were counted. Contacts decreased exponentially from 30.8 per min to 1.8 per min as the inter-prey distance was increased from 1.3 to 10.2 mm. Such a decrease of the number of contacts would be expected if a particle moving at random with a constant speed contacted the eggs. However, by calculating the area 'searched' by such a particle Laing was able to show that the walk of the searching *Trichogramma* female is far from random in that it searches over a wider range when the prey-density decreases (Table 2). It can be seen that the area of search approximates the area of distribution of the eggs more closely when the latter are close together than when they are farther apart. One must conclude that the parasite gets information

Table 2. The relation of the area of search by female *Trichogramma evanescens* (Hymenoptera, Chalcididae) to the area of distribution of host eggs. (Modified from Laing, 1938)

Distance between eggs (mm)	Area of distribution [a] (cm^2)	Area of search (cm^2)	Area of search / Area of distribution
1.3	4.59	7.36	1.6
2.5	9.69	16.60	1.7
3.8	15.24	36.20	2.37
5.1	21.32	46.50	2.18
7.6	34.90	75.50	2.17
10.2	50.40	126.40	2.5

[a] Eggs were distributed over a rectangular surface $(381 + d)$ mm long and $(22.8 + d)$ mm wide, where d = distance between the eggs.

in some unknown way about the density of its host and responds to it adaptively in searching over a wider area when density falls off.

The avian predator is remarkably similar to the insect parasitoid. Carrion crows when presented with quadrate arrays of prey in the form of concealed chicken eggs or camouflaged artificial pastry "larvae" find consistently more prey items in crowded than in scattered populations (Tinbergen et al., 1967; Croze, 1970). In both cases the lower mortality of the scattered population was not due to a lower effort on the part of the predators. "Inter-catch distances" (a measure closely related to giving-up distance) were rather longer the more scattered the prey was (Fig. 19); the difference

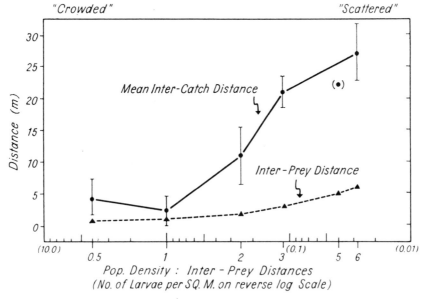

Fig. 19. The inter-catch distance of carrion crows searching for hidden pastry 'larvae' increases more slowly than the inter-prey distance as the larvae populations become less dense. (From Croze, 1970)

between the inter-catch distance slope and the inter-prey distance (reflecting the population density) slope is significant and it demonstrates the power of scattering to multiply the amount of effort the predator must expend to find each prey.

A higher inter-catch distance signifies a longer searching path but not necessarily a higher level of search saturation since Croze (1970) could not tell whether in fact the radius R of the searched area after capture did remain constant, irrespective of prey density; only then would a longer path travelled between two finds mean higher search saturation.

In his detailed analysis J. Smith (1971) gained some insight into the nature of path changes after capture as a function of prey type, prey density and arrangement. Green pastry "larvae" were distributed over a grass lane

51

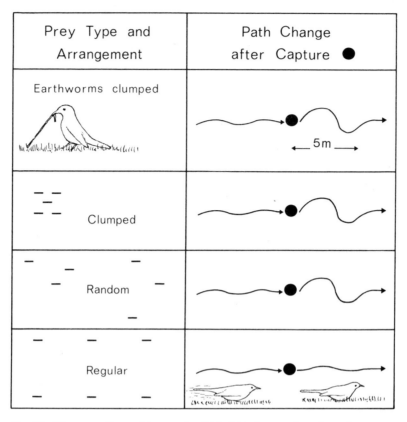

Fig. 20. Changes of searching path of blackbirds as a function of prey type and prey arrangement. In the lower three figures camouflaged artificial prey 'larvae' that were added to natural (underground) prey, caused the path changes illustrated. Density of artificial prey 0.064 per m² on a total grid area of 627 m². For further explanation see the text. (Modified from an original of J. Smith)

at two densities with three arrangements each (Fig. 20). With the high-density arrangement, that is 0.299 larvae per m² on a grid of 183 m², the overall searching path of the blackbirds was so convoluted that any change of path after capture remained necessarily masked; apparently, the birds thereby strived to remain in the prodigiously baited grid area. By contrast, the search path changed strikingly in the low-density prey arrangement. With clumped and with randomly distributed prey of equal numbers the blackbirds altered the sequential patterning of turning after a capture as if confronted with the naturally occurring and moderately clumped (negative binomial index of aggregation $k = 6.93$) earthworms alone (see Fig. 20). (The availability of worms to the thrushes could not be assessed). Accordingly the birds after capture took longer bouts of turns to each side so that they remained in the vicinity of their find (see p. 49). The nature of path changes after capture in the regular prey arrangement of like population size

52

was strikingly different (Fig. 20). After capture the thrushes continued an ongoing path as before but took more time to cover the moves after capture (see also Murton, 1971: *Columba palumbus*); this may have been a reflection of increased time spent at stopping points after capture. This in turn could have allowed more scanning movements, as the number of head movements at each stop is known to increase with growing length of pause. Still another tactic can be seen in the black-bellied plover (*Pluvialis squatarola*) though prey distribution is unknown in this case. Upon a successful capture the plover makes a new attempt to capture a prey after a shorter move than otherwise; in this way this bird too stays in the vicinity of a find (Baker, 1974).

Functional Interpretation [3]. Both changes in behavior by the blackbirds are likely to promote increased searching of the area around the capture point; the path change by concentrated scanning in that area, and the decrease in forward movement after capture by increasing the time for scanning at each stopping point. This would appear to enhance the likelihood of further captures with prey living clumped.

As the birds use a number of alternate methods, adaptiveness, if any, would have to be facultative rather than fixed. It should be noted that judgement upon the nature of the adaptedness refers to the level of description: the birds' responses to varying environments are facultative in terms of achieving a common "goal", i.e. prey capture; but they are fixed in terms of prey arrangements if not chosen at random in each particular one. In order to recognize facultative adaptations one has to relate the effectiveness of searching, for that matter, to the prey arrangements in which it is shown. If the same "goal" is achieved, within limits, by different methods, these would appear to be adaptive. Achievement of goal was inferred by J. Smith (1971), when seeking those relations, from what would have been, from the point of view of technique, the best solution in each particular prey arrangement.

Changes of path or of speed of movement after capture has only been found in the low prey-density arrangements, not in the high ones. At very high densities search time may become unimportant, and hence changes in searching behavior, as compared to "handling time", so that the birds have no "need" to modify their search to make it more efficient. Hence, the disparity between searching in the two situations would appear to be adaptive. Further, a slowed-down forward movement after capture in the low-density regular prey arrangement would increase the chance of finding the next item because it would allow more scanning movements at each stopping point. This argument would only hold fully if the amount of scanning remained constant regardless of speed of movement. There is a suggestion that with decreasing speed, pause length increases at the expense of move

---

[3] The term 'strategy' has been used in so many different contexts of food-getting behavior as to be almost useless. I therefore propose to reserve the term for a hunting "method", "technique", or "tactic" that has been shown to be functionally superior to any alternate behavior of the predator concerned. Then there would be reason to believe that the 'method' concerned has been evolved specifically to cope with a certain prey or a prey situation.

length so that, in fact, scanning may be promoted at lower speeds. However, the birds did not respond adaptively to all changes in food distribution. According to J. Smith, this may be possibly indicative of a fundamental feature of such a flexible behavior system. Since the bird is required to make a "prediction" of the consequence of a range of possible responses to a particular set of prey stimuli, the very nature of the partially unpredictable environment which dictates the existence of flexibility, must mean that the animal's predictions will not always be accurate. In addition, J. Smith's thrushes spent most of the day eating prey other than those put out for the experiments and this may well have influenced the results.

3. Search Behavior after the Disappearance of Prey

When a predator sees a prey disappear it either tries to flush it from cover (e.g. Angell, 1970), keep the place of disappearance under close watch, or, resumes the activity in which it was engaged before spotting the prey. There is a dearth of even anecdotal information as to what factors determine the search for a vanished prey, known in psychology as the "delayed response" problem (review Fletcher, 1965).

In arthropods the persistence of such search depends on a number of factors. As reported by Chandler (1969), the larva of *Syrphus balteatus* (Syrphidae) which is hunting aphids turns around more on a spot where it killed prey, slows down forward movement, and "casts" laterally more effectively. These changes are, within certain limits, proportional to the duration of the previous contact with an aphid of up to 5 min.

The dragonfly larva's (*Aeschna cyanea*) search, after having lost sight of potential prey, depends strikingly on the time of visual exposure. After a brief presentation of a dummy prey stimulus the larva maintains its fixation position, thus looking motionless for up to three minutes where the dummy had disappeared. Dummy stimuli producing the highest rate of strikes during presentations were also followed by a longer retention of the orientation towards the edge where they had disappeared. However, in the individual larva number of strikes and retention time do not correlate well. This suggests that both measures are not controlled in the same way. Moreover, since the most effective stimuli were followed by longer retention of the orientation position it seems reasonable to assume that the "central prey excitation" (see Chap. 1. A.) determines the latter. In which way tracking and striking during stimulus presentation influence retention time remains to be investigated.

The functional value of maintaining fixation seems clear, for the larva surveys that part of the environment where the prey is most likely to reappear or to move again after having stopped movement and thus lost its main characteristic as prey; on reappearance of the prey the larva strikes immediately as there are no preparatory adjustments (Etienne, 1972, 1973). After longer, and procedurally different, possibly more natural prey presentations the larva shows a searching behavior whose function seems to be less clear (Hoppenheit, 1964 b). The larva first maintains the fixation position and then moves backward, turns, moves forward, pauses, etc., thus moving on a spiral-shaped track. While pausing time and again the larva is prepared to strike, as immediately after disappearance of the prey. During this searching and thereafter the larva may fixate and eventually strike at pebbles in lieu of prey. The same may happen if the true prey "vanishes" by stopping to move.

As in the dragonfly larva there is in owls also some suggestion that "central prey excitation" which has been aroused by a prey before it disappears exerts an influence upon the length of retention. A tawny owl (*Strix aluco*) retained memory for meat that had been hidden before its eyes for not more than 20 min but for a dead mouse for at least 30 min (Macura, 1959). This indicates that by such memory tests central prey excitation, and possibly specific hungers, could be profitably explored.

From the foregoing account it is clear that a prey need not disappear physically but need merely "jump" off the perceptual mechanism involved in tracking it. Such "false" disappearance is taken advantage of by a number of insects, which, after being flushed by a predator, settle again quickly on a matching background on which they will often be overlooked by searching eyes (e.g. Sparrowe, 1972, pers. obs.). In which way the visual appearance of the fleeing prey may keep the pursuer "locked up" for detecting it thereafter will become clear below (see Chap. 2. C. I.).

Predator and prey alike share common interests in that prey animals which have lost sight of a predator also carefully try to keep track of him, yet species differ markedly. While tits, buntings, and sparrows, for instance, after a flying predator alarm unhesitatingly resume their activity, carrion crows at times intently watch the place where a goshawk (*Accipiter gentilis*) has disappeared, until it flies off (Löhrl, 1950 a). A comparative approach to the subject appears rewarding.

II. Long-term Area Concentration

As stated already animals show spatial variations in their distribution. In order to exploit a given prey population economically predators would have to concentrate their search onto areas containing prey as compared to areas with little or no prey, though similar in all other respects. Long-lived predators, as for instance vertebrates, would, of course, profit most from long-term area-concentrated search. Yet also hymenopteran parasitoids (Braconidae), for instance, staying right beside a colony of their prey larvae visit them for oviposition for many days in succession (Morris, 1963).

Sticklebacks which had learned that certain compartments of a maze never contained prey avoided them by making less turns than initially (Beukema, 1968). Field observations of other fish species and others of anurans tend to confirm this (Wigley, 1965 quoted by Beukema, 1968; Heusser, 1970, p. 371 in Grzimek, 1971, Vol. V).

Ullrich (1971) reported on a similar behavior in one of his great grey shrikes that had caught a mouse in a bucket. In the 10 min following capture it dived twelve times in succession into the empty bucket and repeatedly mock-captured this way still after 3 and 9 days (see also Cade, 1967; Sulkava, 1964: goshawk). Memory of this sort may even last for many months (Heyder, 1970). Similarly red foxes which had learned to exploit a large gull colony revisited it time and again, thus demonstrating that it may be dangerous for birds to cluster in breeding colonies though the advantages seem to outweigh such inraids (Kruuk, 1964, p. 45; see also Murie, 1944: wolf). The range of conditions conducive to area-concentrated search still remains to be explored. For example, it is unknown whether the perception of prey alone would be a sufficient condition; would, for example, a predator return to a place where it had merely seen a prey disappear (see previous section)?

A detailed knowledge of the home range is known to aid prey animals in their escape from enemies (e. g. Metzgar, 1967; Goodyear, 1973) and predators such as lions (Schaller, 1972, p. 245 seq.), and perhaps tawny owls (Southern, 1954), in their hunting. But there is a suggestion that prey animals such as ducks (Refs. in Curio, 1963) or ungulates (Schaller, 1972, pp. 236, 369) do not make use of their spatial knowledge in forestalling predation by avoiding places where they have been threatened before.

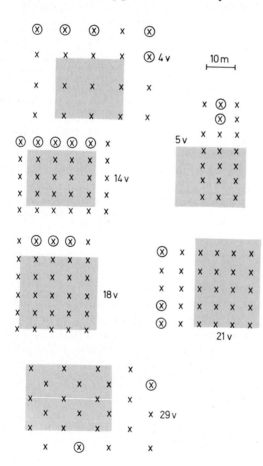

Fig. 21. Long-term area-concentrated search by carrion crows for red mussel shells (*Mytilus*) covering a bait X; surviving shells ⊗. Hatched areas indicate the position of shells prior to the six test days diagrammed; outside these areas shells had never been presented. (From Croze, 1970)

In a systematic way Croze (1970) studied long-term area-concentration of search in carrion crows that had learned to hunt for baited red mussel shells (*Mytilus*) on coastal shingle areas with which they had become intimately familiar. When the crows were offered shell populations which were new in size and shape they searched those areas more successfully in which they had hunted for the red shells before (Fig. 21). The shells outside the familiar areas survived significantly better than those inside. This differential predation is particularly remarkable since the crows did not restrict their movements to the familiar areas but rather landed outside when they flew in from somewhere, or, in between hid a beakful of meat outside the familiar area. Their searching image (Chap. 2. C.) for the red mussel shells operated more efficiently in that area in which it had often been reinforced. At present it remains unknown to which characteristics of the areas the crows were attending, to the pattern of stones or landmarks, or the overall area covered by the prey population. The achievement does not appear to transcend the learning capabilities of corvids (see Schiemann, 1939; Koehler, 1955).

Area-concentrated search of this sort could put a premium on a prey population to move about continually. Perhaps the travelling about of camouflaged insects, more than would be dictated by their food requirements (e.g. Curio, 1965 a), has been brought about at least in part by a selection of this sort.

Long-term concentration need not be directed onto a spatially fixed area, as shown by a marine cleaning-symbiosis. Amongst señoritas (*Oxyjulis californica*) acting as cleaners of other fish there are individuals which confine their cleaning on only a few species while other individuals behave far less specifically (Hobson, 1971). Honeybees similarly partition their food resources amongst each other in that they time and again visit particular small areas of approx. 40 m in diameter, to exploit only certain flowers (*Blütenstetigkeit*), regardless of their particular spatial arrangement (v. Frisch, 1965, p. 252). These observations touch on the problem of the functional significance of individual specialization within a species which will be discussed later (Chap. 5. B. III.).

III. One-prey : One-place Association

From many field observations Croze (1970, p. 55) and his colleagues obtained the impression that crows hunting for food arrived in an area as if they expected to find a particular type of prey there. A critical experiment to test this idea failed to support it but many reasons for this could be adduced so that the idea is far from being disproved, especially since an expectation to find prey in a particular place is an inherent characteristic of the hunting of birds (see previous section).

Honeybees learn that each species of food plant provides nectar only during a certain time of day. A bee that has specialized on, say three flower types, becomes active and visits each of them during their respective peak of giving nectar. During the rest of the day the bee stays in the hive. Thus bees learn to visit certain places at specific times of the day, that is they form apparently the "one-prey : one-place association" (K. v. Frisch, 1965, p. 256 seq.). With the type of food remaining constant they learn to come to up to four different feeding stations at four different times per day (Finke, 1958 in K. v. Frisch, 1965, p. 257). It would be interesting to see whether offering the bees a different food each time would enhance learning performance. Similarly, laboratory rats can learn to visit a particular place in their environment in order to get sodium once they sense the need for it, and to visit another place to get their normal food when they are generally hungry (Krieckhaus and Wolf, 1968). A „central mapping" of a similar kind is borne out by captive chimpanzees (*Pan troglodytes*). When allowed to witness the depositing of up to 18 food items along a track in a familiar enclosure the animals subsequently found most of them. Their search path clearly showed that they remember the positions in relation to each other: they moved from one place to the next so that the distance covered was minimal. Moreover, they searched places with the more desired of two types of food more intensely; and a general area that was baited with three as opposed to an area with two like items, first. Hence, the chimpanzees were able to remember what type of food was hidden in each hiding place and to rank areas in terms of their profitability (Menzel, 1973). Using hand-reared red foxes in the field searching for their caches with voles, Macdonald (pers. comm.) arrived at a similar conclusion.

In view of animals taxonomically so diverse mastering the one-prey : one-place association it would not be surprising to find it established in an insect predator too. The prey approvisioning behavior of certain thread-waisted wasps (Sphegidae) may be regarded as a case in point though on a different scale. Species utilizing several insect species as food for their larvae provide each larva-containing burrow with paralyzed prey items belonging to only one species (Molitor, 1939).

C. Object-concentrated Search

While the search for food has so far been considered a matter of hunting in the right area it will now be examined in how far specific properties of the prey affect search behavior.

I. Existence and Properties of "Searching Image"

1. Ecological Evidence

It was von Uexküll (1934) who originally conceived the idea of a "Searching Image" to circumscribe a perceptual change facilitating the search for a particular object as a consequence of having found it before. Recent work suggests that searching image involves selective attention for certain stimulus objects ("learning to see"), for example camouflaged prey, in preference to others which are accordingly "overlooked". Recently, Krebs (1973 a) listed a number of types of learning which should be excluded if one is to invoke searching image so that actually the usefulness of the concept depends on how exhaustive this list will be. Searching image thus, suffers from being defined in negative terms.

The concept was revived by L. Tinbergen (1960) to account for hunting and prey selection by great tits in Dutch pine forests. Predation on each prey species was assessed by its frequency in the diet of the nestlings, not by the proportion of the population consumed. A specific searching image for a particular type of prey was suggested by the following points: (1) Although suitable prey appears in the environment only gradually, it is often not accepted as food by the birds for several days, but is then accepted abruptly. (2) This delay between appearance and acceptance cannot be due to the birds' rejecting a prey item on first finding it, to differences in the availability of prey species, to preferences for different layers in the wood, nor to differences in feeding technique. Furthermore, novelty was ruled out since the tits readily accepted artificially painted mealworms as prey. (3) The delay cannot be due solely to changes in the conspicuousness of caterpillars by their growing larger, since *Bupalus piniarius* moths upon emerging experienced a comparable delay (Mook *et al.*, 1960).

In his analysis Tinbergen compared the actual percentage predation of each prey species in the birds' diet with the expected predation frequency based on the probability of chance encounters. This led to the important discovery that the percentage predation of four common species of caterpillar prey was considerably less than expected when the prey were at very low densities, increased abruptly and exceeded expectation at moderate densities, and finally declined again below expectation at very high densities (see also Krebs, 1973 a, Fig. 3). Thus predation was relatively most intense at moderate absolute prey densities, though the actual value of these moderate densities may differ for each prey species. Hence the type-3 functional response obtained was in principle the same as that found by Holling (1965) for a deer mouse (*Peromyscus leucopus*) and a shrew (*Sorex cinereus*) hunting for sawfly cocoons (Fig. 22).

The switching at low/moderate densities was ascribed by L. Tinbergen (1960) and Holling (1965) to the vertebrate predator having learned properties of the prey which it hence exploits more economically. The tits were thought to adopt a searching image by random encounter and to concentrate subsequently upon the respective prey, again in random encounters. This would continue until the overall composition of available prey species shifted again to give the birds an incentive to form another searching image. The switching at the moderate/higher densities was ascribed to the predator's tendency to keep its diet varied or because of other reasons (Chap. 1. C. I.). Great tits in fact seldom take any one prey species in excess of 50% of the total diet fed to the nestlings (L. Tinbergen, 1960; Royama, 1970 a).

Furthermore, there is circumstancial evidence in support of the notion of prey-specific hungers in predators which would mean that they would rarely be satisfied with prey of just one type (Chap. 1. C.).

As pointed out by Krebs (1973 a) the ascending limb of the sigmoid curve can only occur if more than one prey type is available; the predator could concentrate its attack on one type only after its density had risen to a level permissive to do so. With only one prey type the curve would simply flatten off at higher prey densities as in a "type-2 functional response" (see Fig. 27); here the predator could not afford to "give away" part of its food

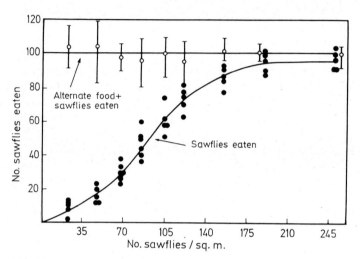

Fig. 22. Effect of prey density, sawfly cocoons, on number of prey eaten by a deer mouse in a homogeneous cage environment. Alternate food was dog biscuit the amount of which has been converted into 'sawfly-units'. (After Holling, 1965)

at low densities. It is not clear whether a predator, confronted with many prey types all fluctuating in density more or less independently and all with different spatial distributions, will show the functional responses described here for a predator hunting in a simple environment with a poor fauna. With this reservation in mind it is gratifying to see that a type-3 response has in fact been found for wild great tits (see also Chap. 2. C. IV. 1.).

Another, though not very strong (see next section) support for the idea of searching image comes from the fact that prey of one type is sometimes taken in "runs" which can hardly be explained on the basis of random selection. For instance, herring larvae surrounded by diverse planktonic prey, feed for a while exclusively on nauplii, then on copepodits, thereafter on *Artemia* eggs etc., though the availability of these prey items had remained the same (Fig. 23). It has been ascertained that "runs" of this sort cannot be attributed to types of prey occurring in patches. Similar runs of one type of prey have also been found in great tits by Royama (1970 a), but were interpreted quite differently (see Chap. 2. C. IV. 1.).

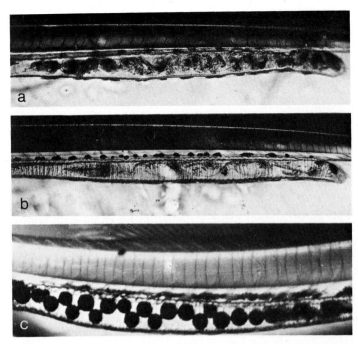

Fig. 23 a – c. Prey of one type each taken in "runs" by herring larvae of 13 – 16 mm body length taken simultaneously out of a tank. Gut containing (a) 21 nauplii, (b) copepodits plus only 1 nauplius, (c) *Artemia* eggs plus 3 nauplii (*at the right end*). (From Rosenthal, 1969 a)

2. *Experimental Evidence*

Studies of Reactive Distance. An improvement in prey detection as a result of previous experience with that prey has been revealed by Beukema (1968). As can be seen from Fig. 24 the probability P that a stickleback searching for food in a large tank subdivided into 18 "cells" will detect a *Drosophila* larva rapidly decreases with increasing distance from the prey. On the other hand, for one and the same distance P increases rapidly with growing familiarization with the larvae, for the distance between successive pairs of curves in Fig. 24 declines in the order *A* through *E*; the sticklebacks reached top detection performance after approx. fifty experiences under the conditions of the experiment.

Tubifex worms as a prey are more appreciated by sticklebacks. Their being detected at much greater distances during the period that *Drosophila* was present, cannot be solely ascribed to their higher attractiveness since also sticklebacks which liked both prey animals equally discovered *Tubifex* at persistently greater distances, apparently because of their larger size and their more vigorous movements. Hence the direct detection distance must, aside from learning, also depend upon the conspicuousness of the prey.

An improvement of prey detection has also been found in rainbow trout (*Salmo gairdneri*), a species that is known from nature to concentrate upon only one or two types of prey at a time. Prey consisted of equal-sized pieces of chicken liver. Average direct detection distance *RD* is given by the formula (Ware, 1971):

$$RD = RD_{max}\,(1 - e^{-a}(E)^{+b})$$

in which

RD_{max} = the distance from which a conditioned animal will attack the prey,
a and *b* = constants,
E = a specified level of experience with the prey.

This equation only holds after the fish have overcome a "latent phase" of some four days during which they do not approach any novel food at all. The improvement of prey detection by stickleback and trout, as measured by an increase of *RD*, might be taken as evidence of searching image formation. Further experiments of Ware (1971), however, seem to show that improvement of detection was due to a prey preference becoming stronger and

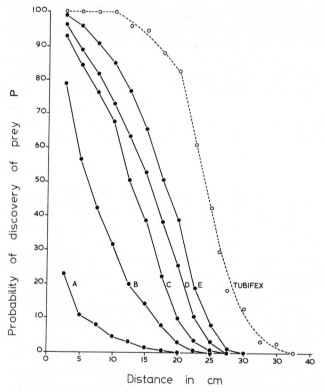

Fig. 24. The probability *P* that a stickleback will discover a *Drosophila* larva (*A* through *E*) or a *Tubifex* worm, as a function of the direct detection distance and of learning. *P* is calculated as the fraction of all encounters in which 9 sticklebacks discovered the prey. *A* through *E* represent successive periods of 20 eating experiences each. (From Beukema, 1968)

stronger. The tank in which trout were trained to feed had a black floor so that white prey were conspicuous and black prey cryptic. Trout trained on white prey and then switched to black took about as long to reach maximum detection distance as formerly with white prey (although the level of performance remained significantly lower). Since the white prey were easily visible from the start the change in detection distance did not seem to be a function of the ability of the fish to see the prey. Hence, neither Beukema nor Ware have demonstrated searching image as defined above. It should be stressed, however, that the difference between "learning to see" and "forming a preference for familiar food" depends on how "learning to see" is defined in operational terms.

Herring larvae detect their prey at ever increasing distances as they grow. Such ontogenetic changes have been solely ascribed to growth processes in the retina (Rosenthal, 1969 b; Rosenthal and Hempel, 1970). From the foregoing account of the search in adult fish, however, it will have become clear that an improvement of prey detection as a result of individual learning could explain the observation in herring larvae equally well.

The findings on fish reported upon have an important repercussion upon Holling's (1966) "experimental component analysis" of feeding by predators. In the praying mantis on which the model was based, reactive distance changes as a function of hunger, but not in stickleback and trout in which it changes directly in dependence on previous experience with particular prey stimuli. Accordingly Holling's model of feeding behavior would have to be modified to represent predation by a vertebrate too. On the other hand, the findings in fish ought to instigate experiments in invertebrates on the possible influence of learning upon prey detection along the lines indicated.

A Rigorous Experiment. The best evidence for searching image to occur and to facilitate prey capture has been adduced by Dawkins (1971 a, b; see also den Boer, 1971: *Parus* sp.) in her experiments with domestic chicks in which previous experience was accurately controlled. The chicks had to search for grains of rice which, by virtue of their green or orange color, were conspicuous or "cryptic", i.e. matched the color of the background. Stones of various sizes glued to a hardboard 60 × 55 cm formed the "background" which was either of green or orange color.

Chicks were fed for three weeks with either green or orange grains of rice on white background and then tested for their ability to see either cryptic grains of one color on a like-colored background or conspicuous grains of both colors on the same background in another experiment. Though the chicks, on average, took grains in all four types of experiment more slowly in the beginning they improved in picking up and swallowing them much more rapidly in the conspicuous grain test; in a run of 100 grains it took them 9.1 sec to find the first five of the conspicuous grains but 240.0 sec to find the first five of the cryptic grains. To find the last five grains the chicks needed approximately equal spells of search time, i.e. 3.0 or 3.6 sec.

From this one may conclude that at least part of the change in the speed of finding cryptic grains was due to a change in their ability to detect those grains, since changes due to other factors such as fear of novelty, hunger, or ability to handle rice could be expected to affect the birds equally in all tests. The change leading to the improvement of detecting cryptic food was of a more central nature (see also Croze, 1970; p. 42) since peripheral adjustment of vision by altering head or eye movements did not seem to take place: The search posture with head lowered below the top of the back was much longer in the cryptic grain test than in the conspicuous grain test before picking up their first grain. This again suggests that they did not see the cryptic grains at first although they were looking at the ground.

In further experiments Dawkins (1971 b) demonstrated that the chicks "attended" to color cues, in a broad sense, rather than to non-color cues when sampling conspicuous grains, whereas they did the reverse when sampling cryptic grains which by the very nature of their crypticity would, of course, not offer color cues by which they could be profitably detected. However, the results of diverse experiments bear out that the chicks' "attendance" in fact depends on the precise searching situation. A result favoring the view of the chicks attending to color cues when sampling conspicuous grains was obtained with the set-up illustrated in Fig. 25. The chicks began sampling grains in the arena mentioned above. After having eaten 10 grains, a test card with two conspicuous and two cryptic grains was quietly laid into the centre of the arena. After sampling conspicuous orange grains (Fig. 25 a) chicks tended to peck significantly first at one of the conspicuous green grains (arrow in Fig. 25 a) of the test card, while when the same

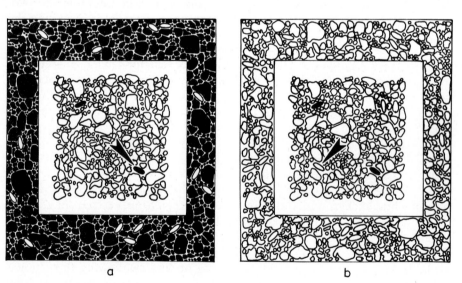

a   b

Fig. 25 a and b. Choice between two cryptic orange grains (*top right and bottom left of central test square*) and two conspicuous green grains, amidst conspicuous orange 'sample' grains on green background (*a*) and amidst cryptic orange 'sample' grains (*b*). (From Dawkins, 1971 b) chicks had been sampling cryptic orange grains delivered their first peck at one of the two cryptic orange grains (Fig. 25 b). It was further shown that the chicks, when sampling the conspicuous grains, were not attending to the particular orange-green contrast as they responded to the reversal of it subsequently; they tended rather to respond to a color contrast as such, thus confirming the view that during their previous sampling they were attending to color cues in general. For the results obtained it made *no* difference whether the chicks were familiar with the color of the conspicuous grains or not.

Experiments in which the chicks were given a choice of the same sort of grain but on two different backgrounds, one matching the color of the grains and the other not, did not fulfill the prediction above. Chicks that were presented with a choice of two cryptic and two conspicuous orange grains on the respective background (Fig. 26), invariably pecked the conspicuous grains

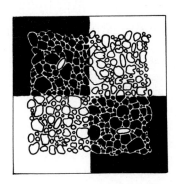

Fig. 26. Choice between two cryptic orange grains (*bottom left and top right*) and two conspicuous orange grains on a green background. (From Dawkins, 1971 b)

first, regardless of their sampling the conspicuous or cryptic grains before. It seems as if the chicks are making use of a searching image only in a situation with a uniform background; otherwise they tend to select first as a background one which offers no difficulty in detecting what is sought.

The experiments with a uniform background on which to search, are compatible with the idea that chicks become able to see cryptic grains better after having just eaten them, as compared to having just eaten conspicuous ones. This could be explained by "shifting attention" from color-cues to non-color cues which enable them to detect cryptic grains (Dawkins, 1971 b). The results forcefully support the hypothesis that searching performance in the cryptic grain situation was due to a shift of attention as a consequence of sampling cryptic grains before. Hence, we may tentatively, yet operationally define searching image as perceptual change which leads to an improvement of searching performance, as measured by items found; as noted above, this happens regardless of familiarity with the items searched for prior to sampling them.

Rapidity of the Formation of a Searching Image. The carrion crows of Croze (1970) which were used to find a bait under a mussel shell, needed, on average, only 2.3 ± 0.5 experiences to become proficient in searching shells of a particular color. The need for just more than one piece in a sample could explain the lag between emergence and acceptance of a prey species observed by Tinbergen (1960) in his tits (see Chap. 2. C. I. 1.). On the other hand, caged European jays and chaffinches exposed to camouflaged geometrid twig caterpillars learned quickly, even after only one experience of treading upon a larva by chance, to find others or similar looking pieces of twig (de Ruiter, 1952) (see also p. 74).

Retention of Searching Image. The crows retained their improved searching for a particular type of shell from one day to the next (see also Dawkins, 1971 a) and sometimes over many days (see also Murton, 1971). After mussel shells were left "blank" (i.e. unrewarded) for eight days the number of shells turned by the crows stayed at first at almost 100% and then rapidly declined. When, after that period, a single shell amongst the whole population was discovered again by a crow to be baited, the proportion of shells turned "leapt up" to approx. 60%, i. e. the crow turned another 30 shells. From such behavior Croze (1970, p. 36) concludes: "The crows tend to divide their daily feeding routine over many different areas, each of which is probably associated with a certain kind of food . . ." (see Chap. 2. B. II., III.).". . . If the supply of prey in a particular area is seasonal, it would be advantageous to "remember" what the prey looked like. The next season, then, just one or a few prey could initiate a concentrated search." It must be emphasized, however, that retention of a searching image over such long spells of time is distinctly unproven.

The rate of forgetting is influenced by other items eaten in the time between finding two cryptic prey. Dawkins' chicks which had just been eating conspicuous orange grains on green background were more likely to forget how to detect cryptic grains, even if the cryptic grains were also orange, i.e. offered on an orange background.

The crows proved exceedingly quick in switching from one type of shell to another (mussel vs. cockle) within a day after the shell that had proved rewarding the day before was found blank and the previously blank shell was found to be rewarding. If this is taken to mean that the crows were forming a new searching image as quickly, and if it is remembered that the old one is retained for many days (see previous paragraph), the conclusion seems inescapable that the crows have at their disposal more than one searching image at a time though it must be acknowledged that one may be dominating at each time.

Human subjects required to search for a letter within many rows of meaningless ones perform, after sufficient training, equally well regardless of whether they are searching for 1 or 10 letters simultaneously (Neisser, 1964), and much higher numbers of items searched for simultaneously are likely (Neisser, 1966). Similarly, when subjects were asked to detect a numeral hidden within lines of letters, an advance knowledge of the numeral made no difference as to the quickness of search performance (Sperling *et. al.,* 1971).

A phrasing of searching image in terms of "shift of attention" (Dawkins, 1971 a, b) implies that the subject pays more attention to certain items than to others which are thought to be "overlooked". As reported on by human subjects they do not see meaningless items while searching for the relevant one(s) and that everything is "blurred" except the item sought which suddenly stands out clearly against an "amorphous" background (Neisser, 1966). This would marshal the hypothesis that searching image involves a perceptual change that is central rather than retinal (see also Alcock, 1973 a). At present, however, it cannot be ruled out with confidence that an animal, upon adopting a searching image, sees other food items and merely does not respond to them. The human subjects mentioned had to search with the utmost quickness whereas a predator travelling over its hunting ground may well be capable of similar achievements but will presumably not utilize them because its speed of progression, and hence surveying new ground, is of much smaller magnitude. Yet the "runs" of prey of one type, as observed in fish and birds, are difficult to account for without assuming a certain degree of "single-mindedness", if only temporarily. The simultaneous existence of several searching images (see above) would suggest that the "single-mindedness" is a relative one.

Specificity of Searching Image. When crows were offered arrays of a shell familiar to them with an equal number of shells differing from the familiar one, it turned out that they had attended to color, form, and structure. Furthermore, specificity of the stimulus sought increased with growing experience, thus saving the predator more and more time which initially he had to waste by testing similar but inedible objects (see also de Ruiter, 1952). Thus initial generalization gave way to more and more perfect discrimination. This accords well with psychologic findings about an ever-sharpening peak of generalization gradients as a result of the subjects having more discrimination trials (refs. in Croze 1970, p. 53). The initial generalization from the prey discovered to similar but inedible objects seems to be widespread as it occurs in fish (Beukema, 1968; Ware, 1971) and numerous birds aside from those mentioned (Carpenter, 1942).

II. Social Facilitation of Searching Image Formation

A prey that is captured, handled and eventually eaten is, in general, attractive. Observations of those activities by another predator may tell him where to search and/or what to search for. A searching image that arises from watching companions has been suggested for the woodpigeon in nature (Murton, 1971; Murton *et al.*, 1971). If the gregariously feeding birds are given two types of seed of almost equal attractiveness, e.g. beans and peas, but of differing density, the majority of birds come to search for the higher-density food. This means a lower pacing rate than for the minority birds which, by chance, came to search for the lower-density food. If the latter birds would stick to their type of food they would rapidly become separated from the majority of the flock. This rarely happens, presumably because of such birds becoming increasingly exposed to predation (see Chap. 4. B.) and feeding uneconomically. Instead, the "minority" intently watch their fellow flock members and switch from their uneconomic source of food to the more profitable one. In addition, the "observer" birds tend to adjust their pecking rate to that of the majority, even though they may not yet have switched to the most economic food. In order to profit from the majority the members still sampling unprofitably, position themselves at the front of the flock where they are thought to watch their fellow members' feeding responses most profitably. As suggested by Murton (1971) such a majority system can operate successfully only in coarse-grained habitats where the most profitable type of food does not change rapidly with time.

Related to this problem is that of a social influence upon food preferences. Day-old chicks were allowed to watch a mechanical dummy "hen" pecking at either green or orange wheat grains, and were at the same time offered a line of alternating green and orange grains. The chicks responded to the "hen's" showing of sample grains and pecked more green when she pecked green and more orange when she indicated orange. Juvenile chaffinches and house sparrows likewise learn the type of food to take, even though unpleasant, from observing taking of that food by adults. Proximity between the "trainer's" food and that of the "observer" tended to enhance learning success (Turner, 1965). This would explain why fledglings of cardueline and geospizine finches closely watch their parents mandibulating flower heads, dehusking of seeds, etc. (pers. obs.); if young born in captivity, and, after gaining independence from their parents, are released into the wild, they would die because they do not know where and how to look for food properly. In teaching them to fend for themselves the adults provide information not only about the type of food but presumably also about the type of place in which to seek it (see "copying", Chap. 2. C. IV. 1.).

III. Searching Image and "Training Bias"

When given a choice between familiar and novel food, many animals select the type of food that they have eaten previously; when thus eating it in greater proportion they are said to have developed a "training bias" for it (review Bryan, 1973). "Training bias" by its very nature operates against deliberate "switching" between different prey and hence makes the diet become monotonous (see Chap. 1. C. II.). By "switching", a predator changes its behavior to exert disproportionately heavy predation on the prey of greatest relative abundance (Murdoch, 1969; 1971). The balance between training bias and switching may strikingly depend on type of food and feeding schedule. As an example, groups of laboratory mice were given varying amounts of two foods differing only by smell. Training mice a single day with excess food A induced a preference for food B with the preference becoming more pronounced with further training. On the other hand, a single day of training with excess of B induced a preference for B, two days training led to no preference, and three or more days' training induced a preference for A (Soane and Clarke, 1973). Another ancillary factor may be initial preference: if it is very strong, no switching occurs as a result of training (Murdoch, 1969).

If "training bias" operates in nature as well, it might have important consequences for searching prey. Any inference, however, about such a possible influence of training bias is hampered by the fact that, as a rule, prey in nature is encountered one at a time, while captive animals given a choice between two foods do have easy acess to both simultaneously. However, simultaneity may then be more apparent than real since trout detect familiar food from farther away than unfamililar food so that two types of food, despite their being offered simultaneously, are not seen with equal ease (Ware, 1972); this may continue even when both foods are superabundant. Thus even with an abundance of food as prevailing in a choice situation the animal may face a task which is related to searching in nature. However, as argued by Bryan (1973), training bias differs from searching image in two ways. (1) Rainbow trout need at least nine days of training with one type of food (e.g. *Tubifex* worms) in order to exhibit training bias for it. It is difficult to see, however, why this period should appreciably differ from the amount of experience needed to develop a searching image (see L. Tinbergen, 1960; Mook *et al.*, 1960) although only very few experiences may suffice (Croze, 1970). (2) After being trained upon one food, trout, when given a choice between it and a novel food, eat progressively more of the novel food but still continue to select the familiar food even after many encounters with the novel food. Since with increasing satiation relatively more of a preferred food will be eaten, the familiar food was apparently less rewarding for the trout. But crows with a searching image behaved similarly (Croze, 1970). Although they would readily switch from a blank type of shell to a rewarded one they would also continue to turn over blank shells despite their misses each time. Though the two situations differ, of course, by many details, the relation of the amount of reward may be similar. The argument overlooks that even an extremely low rate of reward may be reinforcing and therefore have similar effects as a little preferred food. Thus the arguments presented by Bryan do not seem to refute the idea that a kind a searching image may be involved and in fact it is not clear how the two concepts differ fundamentally.

The situation may be different with imprinting in which animals prefer the food on which they have fed early in their life (see Chap. 3. F. II.). However, primacy does not always dominate over recency, as would be the case for imprinting in a narrow sense (e.g. Hess, 1959); when given a choice, rainbow trout select the more recent of two foods one of which has been the first in their life. As demonstrated by controls this effect results from conditioning to the second food and not just from forgetting the first (Bryan, 1973). Results such as these would permit distinguishing operationally between food imprinting and training bias and further work would have to show whether the distinction is a useful one.

Recent studies suggest that training bias may also involve a preference for a certain locality of the environment rather than for a certain prey (Chap. 5. B. III.). In this case searching can only be invoked for the initial immigration into that locality. Once there, a different sort of behavior must keep the animal in place.

IV. Searching Image and Profitability of Hunting

1. Ecological Evidence for Profitability of Hunting

General. An Organism is thought to allocate its resources of time and energy in the most profitable or "optimal" manner. Optimization models have been proposed that usually consider how a predator maximizes net energy yield per unit hunting time per unit energy requirement (Schoener, 1969 a, b, 1971; Rapport, 1971; MacArthur, 1972). The predictions made by generalized optimal foraging models are on the whole too simplistic to provide much guidance in studying behavioral mechanisms (Krebs, 1973 a). It is important to note, however, that a model on the optimal use of "patches" (= parts of the species niche) forms an alternative explanation to searching image as deduced from observations in the wild. The model assumes that the predator maximizes its average rate of prey capture, and is capable of ranking prey patches (and types) in terms of their profitability (see e.g. MacArthur, 1972). The assumption needs therefore to be examined.

Optimal Use of Patches. The sigmoid functional response curve which had originally given rise to the idea of searching image has been interpreted by Royama (1970 a) in quite a different way than by L. Tinbergen (1960), Holling (1965), and others. Although the concept of searching image as conceived by L. Tinbergen has been established by subsequent experimentation (Chap. 2. C. I. 2.), including titmice (den Boer, 1971), he may have been right for the wrong reason. Similarly to L. Tinbergen and his associates, Royama (1970 a) studied the composition of the nestlings' diet of great tits in mixed woods in Japan and near Oxford. After the prey species had been identified and hence the food-plant pin-pointed where they had been collected, one knew in which "patch" of the hunting territory the tits had found each particular food item. Certain features of the tits' predation are thought to be at variance with the searching image hypothesis: (1) Prey species began to appear in the diet also after some delay, but, more important, were still available when the tits had given up searching for them. This would not be expected if searching images which should be continuously reinforced would have governed prey selection. (2) Parents were eating species different from those brought to the young. (3) Some very common and palatable prey were not taken in large numbers. This point is a weak one because at extremely high prey-densities the need for variety might mask an existing searching image.

Direct observation suggested that the tits were exploiting feeding patches rather than prey species. For instance, from the preponderance of certain larvae in the nestlings diet it could be concluded that the tits concentrated their hunting on oaks entirely, searching the trunks and not the foliage. Niche exploitation of this sort led Royama to conceive the concept of "profitability of hunting" according to which the predator allots its searching time to the most rewarding patches of its territory as to maximize its yield. As will be recalled, exactly the same performance has been demonstrated in bees, rats, and chimpanzees which, additionally, also associate the quality of a food with a given place (Chap. 2. B. III.).

Theoretical considerations led to a further important insight. As is well known, some predators respond to increasing levels of prey-density with an ever-diminishing rate of increase, i.e. the number of captures per unit time invariably reaches a plateau above a certain level of density. An example is the "type-2 functional response" of individual female mantids to different levels of housefly (*Musca domestica*) density after 36 h of fasting (Fig. 27).

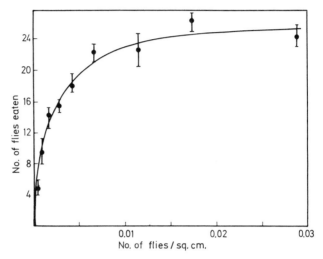

Fig. 27. Type-2 functional response of individual mantids (*Hierodula crassa*) to density of adult female houseflies in a cage of 1.19×0.58 m. *Each point:* average of three replicates; vertical lines: ranges. (From Holling, 1965)

Prey density had been kept constant during each 8-h cycle of observation (Holling, 1965). The same curve in principle has been obtained for other animals as well (refs. in Royama, 1970 a). The unanimity suggests that the trend is fundamental.

To find out if the compression of available time can make the curve gradually flatten out Holling (1959) designed a simple model experiment. A blindfolded girl was asked to dab with a finger and pick up sand paper discs scattered over a table top at varying densities, each time at the highest sampling rate possible. With increasing density of the "prey" discs the sampling efficiency flattened out in the same way as the prey-density predation curve though the handling time h remained virtually constant. Hence there can be no doubt that a predator has less and less time available for capture per item the higher prey-density becomes (though at least in the mantid, satiation contributes to the shape of the functional type-2 response, too). Note that even with $h=0$ the curve would have its characteristic shape since at least capture and ingestion will take time.

Following Holling (1965), Royama (1970 a) showed that the number N of prey animals captured is given by the simple "disc-equation" (for obvious reasons):

$$N = \frac{a \cdot D \cdot T}{1 + a \cdot h \cdot D}$$

where

a = A proportionality factor related in some way to the predator's ability to find prey,
D = Density of prey,
T = Total time spent hunting,
h = Time spent handling each victim (i.e. responses following capture, e.g. killing, preparing, eating, digesting, carry to nest).

The "disc equation" holds on the assumption that prey density remains constant, a not too unrealistic premise in regard to insectivorous birds in the wild. The disc equation suggests that the number of prey taken per unit time increases as the prey density increases but gradually levels off to an asymptote with $N = T/h$.

If different prey species live in different "patches" of the predator's hunting range, as shown for several prey of the great tit (see above), the problem for the hunter is to allocate its time between them in the most productive way. The predator has to assess the profitability P of its hunting for a prey species in its patch. P may be assessed by the hunter in terms of biomass or nutritional value etc. If it can freely move among different patches, and if prey-density is crucial, it will maximize prey capture it if hunts during each time T in the most proficient one. As can be easily derived from the basic form of functional response curves (Figs. 22, 27) the most profitable allocation of time between patches becomes increasingly less important as prey-densities rise to high values. In this way aggregations of predators in places with prey-densities permitting maximum yield (e.g. Pitelka et al. 1955; Goss-Custard, 1970; further refs. in Royama, 1970 b), carried to an extreme in the nomadism of whole segments of a species (e.g. Lack, 1954), can now be analyzed more accurately in terms of patch hunting.

Now we have arrived at a position where the number N of captures can be assessed as a function of prey-density D. If it is accepted that a predator hunting in a patch has increasingly less time T available as D rises (see above), three hypothetical curves A to C may be envisaged (Fig. 28 a). From these curves values of T as a function of D are obtained and values of N calculated from the disc equation. From this operation the relationship of N to D is developed (Fig. 28 b). As thus shown by Royama, regardless which curve from (a) is adopted the resulting curves in (b) are invariably sigmoid (also elaborated mathematically by Royama in a letter, 1972, to Murdoch). If N is plotted as the percentage of all prey animals taken which are put in a single category whose density does not vary, as assumed by Tinbergen, the trend that he observed in his titmice is clearly apparent. Thus, without making assumptions about searching images, the same conclusion about the yield of hunting is arrived at as by Tinbergen's actual measurements.

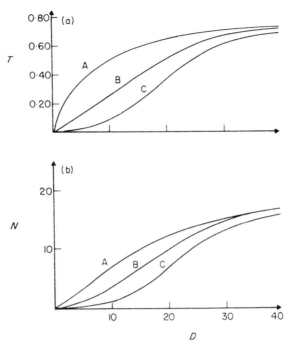

Fig. 28. (a) Some hypothetical trends of the time T spent hunting by a predator in accordance with the prey-density D in a given patch of its hunting range. T scaled as a proportion of the total hunting time spent in all patches involved. (b) The trends of the number N of the prey species taken from the patches concerned, calculated from the disc equation in which the values of T are determined from the curves in (a). (From Royama, 1970 a)

In accordance with these theoretical considerations (for a fuller treatment see Royama, 1970 b), Dunn (1973 a) found that roseate terns (*Sterna dougallii*) robbing fish from other terns respond to the density of their fish-carrying "hosts" in the way predicted by a sigmoid predation curve. Since roseate terns are fish hunters on their own it is difficult to see that they would need to form a searching image for detecting fish carried by other terns. Rather, the properties of Royama's simple model seem to hold, i. e. the compression of hunting time by robbing plus handling time as prey-density increases.

To support his model, Royama pointed to the fact that his great tits carried to the nest one and the same prey species in "runs" that are of a length that cannot be expected on grounds of the random encounter hypothesis of Tinbergen, but which are in line with the profitability concept; an adoption of searching images would have altered the proportion of the prey taken rather than the time sequence of its appearance in the nestlings' diet. However, herring larvae clearly demonstrate that "runs" of the same prey may occur in a homogeneous environment (see p. 60), and Dawkins (1971 a) has shown that searching image formation will lead to such runs. While such observations are undoubtedly correct, they only cast doubt on Tinbergen's null hypothesis of random encounter, but not on his searching image hypothesis to explain the observed data. Moreover, Tinbergen was aware of temporary patch preferences but, as pointed out by Croze (1970, p. 31), such differences would be much more important in the diverse vegetation of Wytham Wood where Royama worked than in the homogeneous pine forests worked in by L. Tinbergen and his associates.

Another attack by Royama upon the adoption of a searching image is equally unconvincing: "... even a *single experience* with one type or prey being sufficient to condition the predator to it" (p. 648, italics mine; see also Krebs, MacRoberts, and Cullen, 1972). This fact in itself is difficult to reconcile with Royama's profitability hypothesis as the only hunting strategy of titmice, for every estimate of profitability would require sampling more than *one* prey item. Theoretically, the bird might use the time until the discovery of the first prey item as an indicator of prey density in a particular patch; however, such a mechanism would be most liable to sampling error. The initiation of hunting for a particular prey after one experience with it would clearly be compatible with, though no proof of, a searching image for that prey (Croze, 1970). Alternatively one would have to assume a further mechanism that conditions the predator to a place after one find therein, a mechanism rather similar to area-concentrated search. Such a mechanism is perhaps best compared to the "one prey : one place association" (Chap. 2. B. III.) since the birds generalize within one type of patch.

A further flaw of Royama's (for still others see Croze, 1970, p. 30 seq.) concerns the discrepancy at high densities between the observed and expected values on the random-encounter hypothesis (see above). He simply discards the possibility that birds might tend to vary their diet, mainly because they would try to collect as much food as possible per unit time, almost regardless of its composition, a view based upon Lack's (e.g. 1954) functional explanation of clutch-size. Secondly, the discrepancy would be explained by patch-hunting alone, without invoking appetence for a change in diet. It must be stressed, however, that even with a superabundance of highly palatable prey various mammals unmistakably strive to keep their diet varied (see Chap. I. C. II.). Furthermore, crows searching for camouflaged mussel shells (see p. 57) were used to get from them only beef meat for many days. After the shells had been baited with mussel meat and thereafter with liver, the rate of turning shells steeply rose after it had gradually declined: A new incentive led to increased searching (Croze, 1970, p. 37).

Great tits hunting by profitability may be said to expect an optimal yield while sampling prey in the patch that is most rewarding at the time. As suggested by Gibb (1958, 1962, 1966), expectation may pertain to an extremely small segment of the feeding territory, for example, a single pine cone. Gibb was able to relate the density of larvae of the eucosmid moth, *Ernarmonia conicolana,* to their actual proportion taken by blue tits (*Parus caeruleus*) and coal tits in the winter. Predation on the larvae that feed on pine seeds could be measured precisely since *E. conicolana* emerges from a smooth round hole, an ichneumonid parasite from a much smaller hole, and tits that have taken the larva leave a large irregular hole (Fig. 29). Number of larvae per 100 cones varied from year to year, from one pine plantation to the other and from tree to tree. Predation on the larvae by the tits rose with increasing density, from 20 – 30% of the larvae in plots of trees with 10 – 15 larvae per 100 cones, up to 70% in a plot with 65 larvae per 100 cones. This disproportionate attack, combined with the large numbers taken, flattened out the marked variations in the density of larvae in different plantations, which had

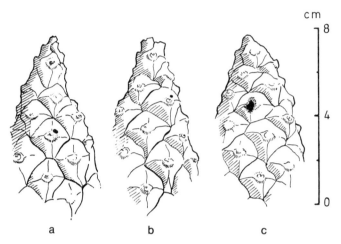

Fig. 29 a – c. Pine cones after parasitism by *E. conicolana*. (a) a normal emergence hole, (b) the smaller hole left by an ichneumonid parasite, *Ephialtes laticeps*, (c) the larger tear left by a predatory blue tit. (After Gibb, 1958 from Lack, 1966)

been present at the start of the winter (Fig. 30). There was evidence that the tits built up an expectation for the profitability of cones: at low densities they sampled the larvae, but rejected them as an uneconomical food on which to concentrate. Above some 10 larvae per 100 cones exploitation increased with increasing density. Conversely, at high densities which were side by side with patches of low densities, the tits sampled less larvae than possible; upon finding the expected number of larvae tits would leave a cone and move to another, even though the first could contain up to 16 larvae. Thus the tits seemed to form an overall expectation by averaging their sampling experiences over more than a small patch of pines and in this way saved time for feeding on a different food. Beyond judging on density per cone, the density of the larvae over a larger area contributed to the level of expectation, for in a winter when cones were scarce the tits still sampled cones density-dependently but at a far lower level, although the cone density of larvae was, if anything, higher (up to 256 per 100 cones) than in the previous year.

Since *conicolana* larvae are concealed beneath the cuticle of cones and the birds have to locate them by tapping their searching image, if any, must be tactile instead of visual.

Gibb's hypothesis of hunting by expectation has been challenged by Tullock (1969) on the grounds of principles of human economy. The disproportionate attack of the tits upon patches of varying larval density (see Fig. 30) could theoretically come about by the tits first concentrating their predation on plots with the highest larval density while others would be depleted only slightly. As the most profitable plots become depleted tits would extend their attack on increasingly less profitable ones until the picture of exploitation as observed by Gibb would obtain. This explanation would render it unnecessary to invoke hunting by expectation as outlined above, and is,

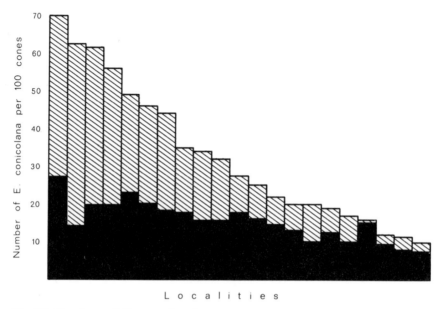

Fig. 30. Numbers of *E. conicolana* larvae per 100 cones before (*full columns*) and after (*black*) predation by tits in winter 1955 – 56. Each column represents one locality and they are arranged in descending order of the density of larvae. (From Gibb, 1958)

moreover, empirically testable. Tullock's explanation, however, appears unlikely because this is not the way tits forage; they tend rather to roam over and deplete a large area each day instead of depleting patches at declining profitability (e.g. Hinde, 1952; Gibb pers. comm.).

A more serious criticism has been leveled by Krebs (1973 a) based on recalculations of Gibb's data by Simons and Alcock (1971). They pointed out that if one plots numbers of larvae eaten per cone on larval density, the leveling off of the predation curve is less marked than if one plots percentage eaten, as Gibb did. Hence, the data do not convincingly support the idea that the birds stop searching after finding a fixed number of larvae. Likewise, Krebs (pers. comm.) when working with *Parus atricapillus* found no evidence for expectation of a given number of larvae, not even after a long experience with a uniform density of larvae per cone. This, of course, does not rule out that an optimal use of patches involves a relative ranking according to their profitability.

Optimal Prey Size. Although small prey may be available in large numbers adult great tits concentrate upon large prey, e.g. noctuid larvae, to feed their nestlings (Kabisch, 1965; Royama, 1970 a). Apparently the longer flights to the nest are profitably counterbalanced by larger prey. After fledging the distance between points of capture and the young has shrunk to a fraction of what it was before. Accordingly the parent birds feed the young on the same items that they feed upon themselves (Royama, 1970 a). Increasing the size of single food rations when feeding young (see also Verner 1965; Lind, 1965; Root, 1967; Kniprath, 1969; Tenovuo and Lemmetyinen, 1970; but see Siegfried, 1972) has been explained by Brosset (1971), beside the functional interpretation given, as anti-predator behavior in that feeding visits to the nest are reduced in frequency (see also Skutch, 1949). However, an interpretation in terms of optimal foraging seems to be more appropriate in at least one other case. Roseate terns in mixed colonies steal fish from conspecifics and three other tern species. As discovered by Dunn (1973 a) only fish of intermediate size are stolen though smaller ones could be robbed with greater ease, obviously to minimize number of flights to the brood (see also Hopkins and Wiley, 1972). That still larger fish are not robbed in proportion to host terns carrying them could be explained also in terms of profitability; they would require chases apparently too uneconomic and perhaps futile, since hosts carrying large fish were more in fear of being robbed, so they themselves may asses their quarry's profitability correctly. Also, larger fish may have presented handling difficulties as they are larger than the commonest prey that is obtained by fishing.

In Conclusion. In "hunting by profitability" a predator is visualized as depleting patches of its feeding ground in the order of declining food abundance, beginning with optimal patches first. This concept can better explain than that of searching image, certain characteristics of sampling prey by wild tits and terns. Together with a knowledge of the frequency of captures at a given prey-density, the concept of profitability of hunting leads to a family of functional response curves which are invariably sigmoid. Hence, the concept furnishes a more parsimonious explanation than searching image for the ascending limb of these curves, and furthermore, a more parsimonious explanation than appetence for variety of diet, for the plateau of these curves. On the other hand, by invoking searching image, as experimentally established for captive birds (Chap. 2. C. I. 2.), certain other features of hunting by wild tits can be better understood. Whether birds do have an expectation for a fixed number of prey items in a given patch, rather than rank a patch relative to others, is still not clear. Finally, it is pointed out that the selection of optimal prey size is, at least in one case, best explained by assuming profitability of hunting.

2. Experimental Evidence for Profitability of Hunting

That predators are able to rank food patches according to their profitability is shown by two laboratory studies on songbirds. In one of them (Smith and

Dawkins, 1971) hungry great tits were used to find mealworms (*Tenebrio molitor*), one at a time, in small pots by removing a cap from them. Mealworms were then offered in series of trials in four different densities with 1, 4, 8, or 16 worms scattered randomly over a quadrat array (1.2 m) of 256 pots each, with the four arrays sited in an aviary (3.7×4.6 m). Before tests, the tits visited arrays when still equally rewarded at different rates. In the test, the least-visited area was then rewarded most and vice versa. It took the tits from 1 to 17 trials to concentrate their effort of search on the highest density of sixteen mealworms per 256 pots. While they spent equally little search time in the three least rewarded areas they searched proportionately longer in the high-density area (Fig. 31). The failure to discriminate between the three low-

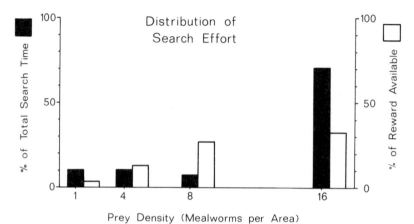

Fig. 31. Distribution of search effort in relation to prey-density over the last eight trials of five great tits in one experiment. *Black bars:* observed distribution of searching (excluding handling time); *empty bars:* proportional distribution of mealworm prey over the four areas. (From Smith and Dawkins, 1971)

est prey-densities is difficult to account for on Royama's profitability hypothesis though the discrimination between the most proficient area and the rest presents a good agreement (see also Simons and Alcock, 1971).

After profitably sampling the total array of pots prey-density was reversed so that arrays 1, 4, 8 and 16 contained 16, 8, 4 and 1 mealworms, respectively. The tits continued to search intensely in the location of the former high-density area for at least four trials after the reversal. Even dramatically high reward rates in the new high-density area could not distract one bird from doing this. This and similar findings demonstrate the effect of previous experience upon ongoing search. The effect of recent reward rates decayed rapidly when it conflicted with longer-term experience. Such behavior is adaptive in that it makes the predator search in formerly rewarding areas and so continually keep track of a new prey potentially emerging. Secondly, the after-effect might explain, why wild tits showed a lag in their search behavior after emergence of a new type of prey. In experiments with red-winged blackbirds (*Agelaius phoeniceus*) Alcock (1973 a) was able to show that the efficiency of food finding was influenced both by the type of food fed upon previously and by the locational cues associated with previous food. The evidence can be taken to mean that searching image and patch hunting may exist side by side.

The conspicuous concentration of the great tits upon the area of highest prey-density may be at variance with other feeding behavior in the wild, in that these birds concentrated their search much more upon moderate than upon high densities (Gibb, 1958). But it should be remembered that the wild tits were mainly nonterritorial birds searching over fairly large areas whereas the captive tits were intimately familiar with a small area and a highly stable, i.e. predictable situation. Moreover, high-density patches of *conicolana* larvae were extremely rare (Gibb, 1958, 1962) as opposed to the experimental situation of Smith and Dawkins, so that the behavior in both situations appears to be adaptive; a second reason to assume this is that animals, as a rule, live aggregated.

A particular variant of hunting by expectation involves social facilitation by conspecific flock members (Krebs, *et al.,* 1972) and is, in this respect, reminiscent of a similar influence upon searching image formation. Captive great tits which have observed a fellow member find a prey, for example, a mealworm underneath a concealing tape glued to a tree, direct their prey-searching not only to the same area or the very site ("local enhancement" in the sense of Thorpe, 1958) but even to the same type of place, thus displaying a faculty called "copying" [4]; one experience sufficed.

A visit of a tit at a place with no reward resulted in less searching at that place. Such unsuccessful visits produced more local enhancement when the tits were used to food which was distributed in clumps rather than dispersed singly. Learning the general area in which food is likely to be found is, in the wild, perhaps useful in locating a tree with a rich crop of seeds. Learning the type of place in which to find food would be advantageous even if the food were more scattered. The adaptive value of "copying" can be inferred, furthermore, from a number of facts concerning the aggregation of certain prey insects and from considering it to be of advantage if a bird that, upon "copying", can learn at least the type of site that may harbor a reward. Recently, Krebs (1973 b) has shown that copying operates also between two chickadee species (*Parus atricapillus, P. rufescens*). Copying was also involved when the two species were trained separately to forage in different positions in experimental trees, they converge in their foraging behavior when they are put in mixed flocks.

Note that copying induces a bird to switch rapidly between types of prey and thus furthers variety of diet, whereas the adoption of a majority decision as reported for woodpigeon feeding is a conservative feeding policy which, however, enhances flock cohesion.

Another qualification pertains to the distinction made above between area-concentrated search and object-concentrated search. The experiments on

---

[4] This term had been previously applied in a wider context to denote true imitation (". . copying a novel or otherwise improbable act . .", Thorpe, 1958, p. 122) or learning from observation which type of food to take (Turner, 1965; Murton, 1971).

copying have made it clear that strictly speaking it is short-term area-concentrated search, though of a special nature. Since in many cases the actual site "copied" will lead to finding a particular type of prey, search may readily become concentrated upon the latter and hence may merge into search that is object-concentrated. Thus the distinction of both types appears to have only limited value.

The ability to copy the movements by which prey can be extracted from a known place is rather poorly developed. An observer bird of one of three species (*Parus atricapillus, Zonotrichia albicollis, Muscivora tyrannus*) who was allowed to watch a conspecific "trainer" as it uncovered mealworms hidden in a tray, failed to uncover the prey by observation alone. When isolated the "observer" performed exploratory movements directed to the tray and eventually uncovered the prey only if it had been rewarded with a mealworm in the presence of the "trainer". Observer birds that had not received a reward merely approached and pecked less at the tray and hence failed. Thus the birds only learned by social facilitation to direct their trial-and-error behavior spatially at the correct place (Alcock, 1969; see also Turner, 1965 for chaffinches and sparrows).

V. Prey-specific Expectation

Hunting by expectation has been shown to involve an assessment of the density of one particular prey species upon sampling it. Quite a different sort of reward expectancy that implies expectation of a whole array of prey consequent upon the experience with a different type of prey has been discovered in our laboratory by Schuler (1974). His caged starlings (*Sturnus vulgaris*) were accustomed to receive in each trial a randomized sequence of inedible models, Batesian mimics, and alternate prey, with the number of each group being kept constant from one trial to the next. If the alternate prey is a highly palatable insect, i.e. a mealworm larva or pupa, the starlings do not discriminate between models and mimics as sharply as they can and so "give away" many palatable prey, i.e. mimics. On the other hand, if the alternate prey is genuinely unattractive, e.g. imagos of *Tenebrio molitor,* the starlings are appreciably more discriminating so that the predator strives to obtain more of the palatable mimics; it thus increases its total yield of insects as compared to nondiscrimination. The disparity of the starlings' behavior in the two situations can be interpreted to mean that they build up an expectation, upon experience with a particular type of alternate prey, of palatable prey coming, either in the form of a mimic or in the form of more alternate prey items. The starlings behave as though they have a certain amount of insect-specific hunger which they try to satisfy during a trial, especially as their standard food was devoid of insects. Similarly, sticklebacks selectively avoid *Tubifex* worms upon having noticed the presence of the more palatable *Enchytraeus* worms, obviously because they expect to encounter more of these (Beukema, 1968; see also Chap. 1. C. I.: prey-switching in hyenas).

What has been termed expectation in the preceding chapters, possibly denotes little more than a miscellany of mechanisms differing widely. Concepts such as searching image, profitability of hunting, reward expectancy, or retention of orientation (see Chap. 2. B. I. 3.) have in common that the predator is more attentive to some stimuli than to others. Yet one has to be aware of the extrapolation beyond actual observation these concepts imply, namely the "existence of mechanisms in the absence of the stimuli which activate those mechanisms, when the evidence depends almost entirely on what happens when the stimuli are present" (Hinde, 1970, p. 124).

VI. Ecological Implications of Searching Image

Adoption of searching image not only economizes energy expenditure by the predator but is also has consequences for potential prey. To counteract hunting by searching image a prey species may either become rare or variable. The extremely widespread occurrence of overt polymorphism [5] in animals and plants in fact seems to be due to selection by searching image though a number of other mechanisms are known which also maintain polymorphisms (e.g. Maynard-Smith, 1970). As first demonstrated by Croze (1970), with wild crows, monomorphic populations of baited mussel shells suffered a higher predation than trimorphic populations of the same size and equal frequency of the three constituent morphs (Fig. 32). Days with mussels of one color only alternated with one or more days with mussels of all three colors. The monomorphic populations survived by 3.5 ± 0.5 shells while the trimorphic populations survived significantly better with 10.5 ± 0.9 shells. In other words, a morph had a 3-fold selective advantage when occurring as one third of a polymorphic population. Accordingly the crows did not effectively form searching images because the density of each morph in the trimorphic situation was too low. The fact that the crows did not take runs of any particular morph supports this idea. It is to be noted that being a member of a polymorphic population improved chances of survival by variability per se which, by a feedback process, enhanced survival for the population as a whole. This led Moment (1962) in a comparable situation to call this sort of selection "reflexive".

There are two possibilities to explain how the crows were hampered when looking for the trimorphic shells. The number of shells found in each test, increased slightly though not significantly, although the time from the first to the last shell remained constant. Perhaps the crows were looking for three things at once (see p. 66) and improved their searching images as the experiment went on, or, they gradually learned to shift attention from color cues to

---

[5] Polymorphism denotes the discontinuous, genetically determined variation of a population whose morphs are too frequent to be accounted for in terms of constantly recurrent mutation (review Ford, 1971). Discontinuous variation often manifests itself as "phase variation" or a form of "polyphenism" (Mayr, 1963, p. 150) when environmental factors determine the pattern of variation.

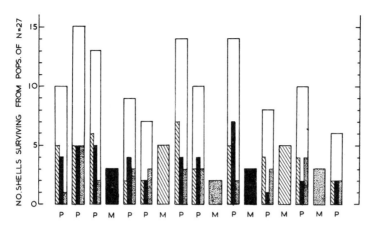

Fig. 32. The number of shells that survived crow predation in monomorphic (M = uniform bars) and polymorphic (P = composite bars) populations. The overall height of the P bars indicates the total number of shells surviving in each polymorphic population; each sub-bar, the number of each morph surviving. Populations tested on successive days with the same pair of crows. *Stripes* = Red mussels, *Black* = Black mussels, *Stippling* = Yellow mussels. (From Croze, 1970)

shape cues since these remained constant throughout the experiment. In the long run, however, it would have been a bad tactic for the crows to start with a single shape-oriented searching image since, when beginning to sample, they never knew what kind of population would await them; even for a monomorphic population attention to form cues alone would certainly be inferior to looking for color cues, for these might be seen at a greater direct detection distance.

Evidence on the survival value of searching image in woodpidgeon feeding behavior is equivocal. Birds that had specialized on a particular item had the same mean food intake as birds which had taken any object they encountered, even though these latter birds automatically had a greater density of food items available. Nonetheless, these birds with the most seeds in their crops were invariably specialists (Murton, 1971).

There are two corollaries from field studies supporting the notion that polymorphism develops in response to predator searching image. (1) Seeds of dove weed are polymorphic with up to six morphs where they are eaten in preference to all other species of seed, by mourning doves (*Zenaidura macroura*), that is in the coastal plains of California. By contrast, dove weed seeds are monomorphic in the Mojave Desert 100 miles away where mourning doves do not occur (Cook *et al.*, 1971). In the same way the color dimorphism of oak saplings, that are either reddish or green, may be explained as a protective device against European jays. These birds locate acorns by searching for year old saplings, the acorn of which is still edible (Bossema, pers. comm.). The interpretation of sapling dimorphism, also obtains support from the fact

82

that no other visually hunting animal seems to exploit acorns in this way. (2) An African land snail (*Limicolaria martensiana*) may be using polymorphism to mask growth in population density. At low density only the most cryptic form occurs. As density increases, more conspicuous forms become more frequent, until, in the most crowded populations, there are four morphs, possibly in response to predator searching images (Owen, 1963, 1965).

If variability confers protection, it is irrelevant how it comes about. Thus for sympatric prey species it would be of advantage to be as dissimilar as possible if living in the hunting range of a common predator. Their similarity could be as detrimental as an increase in numbers of every single one. However, under certain conditions evolution towards dissimilarity or polymorphism may be doomed to be unsuccessful. For instance, any departure from camouflage may be jeopardized, especially if the latter is almost perfect. Selection for camouflage therefore, appears to be incompatible with selection for variability unless a cryptic species developing polymorphism manages to establish each morph in a different patch of the environment, so that it blends again optimally with the background (e.g. Kettlewell, 1955; Curio, 1965 a, 1970 a).

Another condition is supposedly rooted in the need for a varied diet, by which predators may exert selection for arithmetic mimicry (see Chap. 1. C. III.).

Furthermore, predator searching images may have important consequences for the life history of prey. For instance, Blest (1963) found that cryptic, palatable hemileucine moths are rather short-lived as compared to ones that are colored aposematically, and are unpalatable. It is easy to see that a cryptic species, being an attractive prey to insectivorous vertebrates, pays a smaller toll the sooner it disappears, i.e. the earlier it disrupts the development of a searching image. This interpretation would only hold, if relatives of a potential victim, which could give a predator the chance to form a searching image, are on average closer than nonrelatives. This seems a reasonable assumption. Conversely, unpalatable species, whose coloration is designed to function as a warning signal, would gain an advantage from the unpleasant lessons taught to their predators in the past and therefore can afford to live longer.

If the proportions of the different morphs of a polymorphic species are unequal, the outcome of predator selection may be very different as a function of prey-density. Frequency-dependent selection may be "apostatic" in that the predator concentrates upon the commonest morph, so that a rare morph survives relatively better (Clarke, 1962). "Apostatic" selection therefore, maintains variability (Allen and Clarke, 1968; Clarke, 1969; Clarke and O'Donald, 1964), and may operate both via searching image or via prey preference (Chap. 2. C. I. 2.). On the other hand, if the predator hunts for the exception it tends to eliminate the rare form and thus variability (Chap. 4. B.); its selection is "stabilizing". Allen (1972) has shown in field experiments with blackbirds as predators and model prey of two colors (green, brown) that the outcome of selection is different depending upon prey-density. The two color morphs were presented in proportions of 1:9 or 9:1. When baits were at a very high density, selection by the birds tended to be stabilizing, i.e. eliminating the rarer morph, whereas it tended to be apostatic when the density was lower. What constitutes "high", "intermediate", or "low" densities of prey in a given environment will depend on the particular species of predator and prey. For a fuller discussion of frequency-dependent selection the reader is referred to Clarke and O'Donald (1964) and Croze (1970, p. 70 seq.).

Chapter 3

Prey Recognition

Each event in the behavioral sequence search – capture – killing – ingestion of prey may be governed by different kinds of stimuli. For instance, in the bee-hunting digger wasp (*Philanthus triangulum*), approaching, capture, and immobilization of the victim, mostly honey-bees, is steered and elicited by visual, olfactory, and mechanical stimuli, in this order (Tinbergen, 1952; see also Baerends, 1950 and Wolda, 1961: *Notonecta*). Similarly, while prey capture by flagfish (*Jordanella floridae*) is elicited by visual stimuli, swallowing is bound to an intact olfactory epithelium (Foster *et al.*, 1966). The discussion below will be restricted to stimuli eliciting prey capture only. The term recognition will be used throughout when an animal discriminates prey from nonprey or between different prey stimuli, as judged from its behavior.

A. The Stimulus-specificity of Prey Capture

I. Capture-eliciting Prey Stimuli

Prey is recognized by a particular predator by one to three sensory modalities, acting either alone or in combination (Martin, *et al.*, 1974). Every known sensory modality seems to be employed by predators to recognize their prey; recently elasmobranchs were found to detect the biogenic electric field of prey hidden in the sand with the aid of the ampullae of Lorenzini (Kalmijn, 1971).

Most predators are euryphagous and thus may encounter a vast number of prey species which have to be discriminated from nonprey. Because of its ubiquity, body size might clearly serve as a stimulus that characterizes large groups of diverse prey animals. Body size alone, however, can hardly guarantee recognition so that at least one other stimulus, for example movement, must be present. In addition, stimuli other than movement must come into play because almost all predators which are believed to depend upon prey movement are able to detect prey even when motionless. They encompass diverse creatures such as octopuses *(Octopus vulgaris)* (Wodinsky, 1971), young fish (*Coregonus*) (Braum, 1963), skinks (*Eumeces*), anoles (*Anolis*) (Burghardt, 1964), geckos (Morrison, 1937), sand lizards (*Lacerta agilis*) (Svoboda, 1969), and many if not all insectivorous birds (e.g. Kettlewell, 1955; Kabisch, 1965, pers. obs.). Although detailed studies of the stimuli eliciting prey capture are scarce, one may surmise that a great many predators recognize their prey by few stimuli, such as size, movement and/or scent. However, this relative lack of stimulus specificity for prey capture is perhaps more apparent than real.

Stimulus-specifity is most developed in stenophagous predators, the proportion of which varies considerably between taxa. In invertebrates incidence of stenophagy seems relatively high. *Woodruffia metabolica*, a holotrichous ciliate, feeds only upon species of Paramecium (Salt, 1967), and *Stentor sp.*, heterotrichous ciliates, discriminate even between the congeneric Euglenoidina *Phacus triqueter* and *P. longicaudus* (Tartar, 1961). Similarly, parasites confined to only one host respond only to its odor (e.g. Welsh, 1931). In birds the snail kite (*Rosthramus sociabilis*) specializes exclusively on apple snails (*Pomacea sp.*) while the great majority of birds have extremely catholic tastes.

Body Size. If a predator has a choice between prey animals that differ only in body size, it often selects the larger one. This seems to hold true for predators as diverse as opisthobranchs (Connell, 1961), sea lampreys (*Petromyzon marinus*) (Farmer and Beamish, 1973), insectivorous birds (Rothschild, 1963, 1971; L. Tinbergen, 1960; Hamilton and Hamilton, 1965; Root, 1967) and limicoline birds (Goss-Custard, 1970 b). In some cases the smaller of the two prey from the spectrum that the individual species can handle is taken (Kniprath, 1969; Fricke, 1971), or sampling of sizes is at random (Moran and Fishelson, 1971). In some of the social carnivores the problem is complicated, in that packs gauge their size against the size of the intended quarry; thus size selection in this case is a social rather than an individual affair (see Chap. 1. C. I.) When feeding young, many birds take larger prey than they would consume themselves (see Chap. 2. C. IV. 1.). The bias towards larger prey need not always be active selection but may be imposed by the predator's visual capacities. For example, axolotl and rainbow trout catch larger crustaceans, more often than smaller ones, because in both cases larger prey is seen from a longer distance (Maly, 1970; Ware, 1972).

The same bias towards larger prey would be expected if the predator used olfaction; other things being equal the larger of two prey individuals would emanate more odor and thus disclose its presence from further away. This also means that the stronger smelling of two equal-sized prey individuals would be more vulnerable.

Size-specifity is by no means fixed but may be a function of internal factors. In animals as different as mantids and toads, the upper size-threshold declines with increasing satiation (see Chap. 1. A. III. 1.).

Pattern Recognition. In arthropods contour length and solidity of moving prey may (Drees, 1952) or may not (Hoppenheit, 1964 b) influence prey recognition. In order to find out which cues mammalian insect-eaters might use, to detect concealingly colored and motionless insects, Robinson (1970) used captive rufous-naped tamarins (*Saguinus geoffroyi*), highly insectivorous New World primates. They were presented with stick-insects of quite different taxa (Orthoptera, Phasmatodea, Dictyoptera) that were either offered in their natural cryptic position or variously altered. Experiments involved a mantid (*Phyllovates chloropaea*) that was either normal, had its head removed and attached to the abdominal apex, had a head at each end of the body, or was entirely headless. They demonstrated that aspects of the head and configural cues are used in identification. The tamarins could further be "fooled" by small bamboo sticks with an attached acridiid head. When the subjects picked up the artifical prey they lifted it by the stick in the same way as they would pick up a stick-insect by the body behind the head. Moreover, after biting off the head the majority of the animals bit into the inedible stick as well. These observations suggest that the configural cues mentioned must be of a rather general nature and that they become functional only by the presence of the head. Further experiments employed a stick-insect (*Metriotes diocles*) on a matching background of sticks. In the cryptic stick position the legs are pressed close to the body or protracted side by side in line with the long axis of the body thus forming a unit with it. As soon as only one pair of legs was made to stand out from the body, the insect was disclosed to the searching eyes. Similarly, acridiid legs attached to a bamboo stick rendered it so conspicuous that it was captured by half of the subjects involved whereas the legless stick, even when moved, merely released fixation. In conclusion, heads and appendages of insects and configural cues involving these structures are used by the tamarins for prey recognition. To counter this insects have developed a bewildering variety of protective devices that serve to conceal the appendages and/or to disrupt the body contour (reviews Cott, 1957; Robinson, 1966, 1968, 1970).

In order to see to what extent a particular insect pattern would be generalized, rufous-naped tamarins were trained to discriminate between two-di-

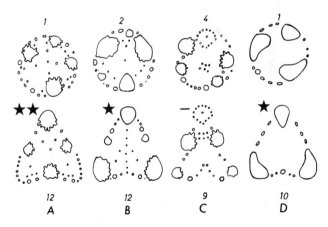

Fig. 33. Discrimination of form-delimiting patterns by rufous-naped tamarins in four choice-situations *A* through *D*. Training was to a moth (rewarded with a morsel of a cricket) versus a circle, a square, or a triangle of equal area, all painted onto cardboard. Testing involved presentation of stimulus pairs *A − D* on an off-white unpatterned background; the black and red pattern elements were different from the ones used during the training. Numbers indicate how often the tamarins chose each respective pattern first; *$P < .05$, ** $< .01$. (Modified from Robinson, 1970)

mensional patterns of a moth, from circle, square, or triangle of equal area (Robinson, 1970). Patterns were painted onto square pieces of cardboard and the moth was the rewarded pattern, with training taking place on an offwhite background. During testing, the tamarins were offered pairs of stimuli consisting of a moth shape and a circle with broken outlines, both stimuli being composed of identical elements. The stimuli on square tracing film were glued onto the background so that three-dimensional cues became minimal. On a background patterned with splotches, resembling the stimulus patterns the tamarins at first failed to discover the moth, though the majority of the capture responses were directed to it. By contrast, on a plain background contrasting with the color of the stimulus patterns, the moth shape scored above the circle, in three out of four stimulus pairs, even if the moth was represented by its broken outline only (Fig. 33). The tamarins had learnt to respond to a generalized moth shape irrespective of how it was composed. When training had begun no bias towards the moth shape had been apparent (Robinson, pers. comm.).

Closer scrutiny of Figure 33 (particularly *B*) reveals that the bilateral symmetry has probably not been the (sole) discriminatory cue which theoretically might serve as a powerful one quite generally. In nature asymmetrical patterning is rare though asymmetrical limb attitudes are adopted by some resting stick-insects and moths (Robinson, 1969 b).

The difference in discrimination performance between both trials with differing backgrounds is likely to be best explained by an effect of camouflage in the trial with a patterned background. This idea is supported by the fact that only on the plain background did the tamarins first attempt to lift the painted pattern.

Knowledge of how previous experience with insect prey may influence prey recognition is badly needed. Unfortunately, there is no control over the learning performance, as a function of reward of biologically relevant patterns (e.g. moth shape), as compared to reward of less relevant patterns (e.g. square).

Movement Recognition. – Some predators respond optimally to prey that follows an irregular path of movement. A dragonfly larva and a water stick-insect (*Ranatra linearis*) prefer prey moving along a zigzag line with the overall direction of progression crossing the predator's field (Etienne, 1969; Cloarec, 1969). Bluegill sunfish (*Lepomis gibbosus*) when tested in a choice situation involving dummy fish moving forward rectilinearly or wriggling, invariably attacked the latter first (Gandolfi et al., 1968). Furthermore, cercariae of a number of flukes (Trematoda), that are fish parasites, mimic the swimming motion of small crustaceans and mosquito larvae, both of which are the staple food of many fish; bouts of upward wriggling and then sinking down alternate with each other (Wickler, 1968). This behavioral mimicry makes sense only if one assumes that fish predators recognize common prey animals (at least) by their particular mode of locomotion. This has been substantiated for common predators of *Daphnia* (Boulet, 1958) and of copepods (Braum, 1963).

88

The predilection of some predators for non-rectilinearly moving prey, poses a problem in relation to protean displays, an anti-predator device of many animals (Chap. 5. A. IV. 3.). Protean displays partly involve locomotion along an unpredictably twisted path and thus resemble a potent signal eliciting prey capture. Of course, one cannot assume that the latter signal is evolved to attract the attention of predators. To reconcile the disparity one has to either assume that a twisted path serves different functions in different animals, or that there are differences between protean displays and other unpredictable paths of locomotion which have yet to be explored.

Speed of locomotion per se may, within certain limits, provide the prey with some protection. Different predators have different preferences for prey speed, so that moving slowly or moving fast makes a prey species escape a certain segment of the predator fauna. For example, guppies (*Poecilia reticulata*) prefer to capture prey (*Diaptomus*) that move rapidly. On the other hand, axolotl (*Ambystoma* sp.) pause some time after the discovery of a *Diaptomus,* and then snatch it up, so that slower-moving forms are easier prey for this predator (Maly, 1970). In this latter case the constraint imposed by prey speed may differ from a true preference, a consideration that also applies to the displacement of prey by water movements (Himstedt and Schaller, 1966). For stalking predators, speed of prey movement determines which out of several hunting tactics will be adopted (p. 138).

Whether movement is required for recognition, may depend on processes associated with prey capture, perhaps to continuing central prey excitation (p. 6). Normally, cuttlefish (*Sepia officinalis*) will only attack prawns (*Leander* sp.) that are alive. Yet, a prawn, taken from a cuttle that has just captured and paralyzed it, will be immediately attacked again if it becomes available even though it is motionless (Messenger, 1968).

A different kind of behavioral change has been found with prey stimuli which when initially perceived in an incounter may "set the stage" for subsequent decoding processes. Juvenile yellow-bellied racers (*Coluber constrictor mormon*), a species of snake, attack a live cricket (*Acheta domesticus*) with a shorter latency than a dead cricket; the latter offers odor alone as a detection cue and is thus located much later. If, however, a live cricket "freezes" and thus ceases to offer movement as stimuli, the snake loses contact with it and has more difficulty in locating it, than it would have with a dead cricket, even at short distances (Herzog and Burghardt, in press). This suggests that prey movement, when perceived by a snake, makes it expect further movement, so that it does not bring olfaction into use. In both cases predators exhibit a short-term shift of attention, the causation of which, of course, may be totally different. The processes involved may be related to the formation of searching image (Chap. 2. C. I. 2.). Moreover, the observation of the snakes suggests yet another conclusion: it demonstrates the survival value of "freezing" for a prey, even after it has been detected.

Some predators require movement of the prey's parts relative to its body. For example, praying mantids (*Parastagmatoptera, Hierodula*) strike most predictably at moving objects with rapid, jerky leg movements and continually mobile wings, though some strikes are also elicited, when all appendages are absent (Rilling *et al.*, 1959). By contrast, tawny owls and long-eared owls (*Asio otus*) do not respond with the complete prey-catching behavior to dead mice or beetles, even if the corpse is artifically moved by a thread. Mice and beetles need to move their legs, or mice their rump muscles (Räber, 1949).

Ontogeny of Prey Recognition. Apart from some higher vertebrates which by some kind of tradition acquire the information of what and how to kill (for examples see Chap. 5. B. III. 3.), the problem of how experience shapes dietary habits remains largely to be explored. Hinde (1958) has proposed that an animal while growing up "learns to take those foods which its repertoire of behavior patterns and structure permit to exploit most efficiently". This view is based on numerous correlations in birds, including intrapopulational ones (Bowman, 1961), between a specific diet on one hand and beak size and shape on the other (review Lack, 1971). From this point of view, a generalized predator would first attempt to catch every potential prey, coming within reach of its sense organs and only after many trials would it know which prey it is able to overpower and which best fits its metabolic needs. The correlations mentioned, however, are also open to yet another interpretation, by which structure and food-getting behavior would be adapted to foods which the species recognizes innately, i.e. independently of individual experience with these foods. The following reasons support such a hypothesis: (1) Numerous insects would presumably be too short-lived to find out by trial-and-error what food is best; this would be especially pertinent to all stenophagous predators and parasites. Similarly, the strike-accuracy of a mantid upon each ecdysis does not change, despite a saltatory change of the body parts involved when striking; the nervous machinery underlying the strike must have changed so as to match exactly the growth of the capture apparatus (Maldonado *et al.*, 1974). (2) Precisely coded innate information is necessary to safeguard a species against attacking a prey that is too big to be carried off or swallowed. For example, ospreys (*Pandion haliaëtus*) were drowned after digging their talons into oversized fish (Glutz *et al.*, 1971, p. 50), and it is only after prolonged training that various birds of prey can be induced to attack large-sized prey that would normally be entirely safe from them, as for instance wolves from golden eagles. Furthermore, a number of vertebrates have also been found, killed by the attempt to attack (Hornocker, 1970) or to swallow, too large a prey (K. v. Frisch, 1959; Burghardt, 1964, refs. in Kniprath, 1969; see also Glutz, *et al.*, 1971, p. 51; Athias, 1972; Zeiss, 1974. The argument is further supported in cases where size is the only stimulus parameter that distinguishes prey from enemies (Ewert, 1970). (3) The recognition of still subtler details would have to safeguard a species from attacking a prey, which itself may retaliate fatally or that it would be imprudent to kill for other reasons (see Chap. 3. A. II.). Visually oriented predators with a large spectrum of highly diverse prey, have been shown to recognize prey by movement and size at least. The question of whether the information about movements is innate in the above sense has been answered in the clawed frog (*Xenopus laevis*).

90

Larvae are planktivores until their metamorphosis. Individuals with no exposure to live prey had their forelegs experimentally freed from the enfolding skin, just prior to finishing metamorphosis. Upon their seeing live prey for the first time, they captured it and pushed it into the mouth with the typical alternating fore-leg movements (Eibl-Eibesfeldt, 1962 a). Thus without any experiential feedback from capture or ingestion, both prey recognition and capture movements were sufficiently developed to render ingestion successful. Furthermore, praying mantids hand-raised on a diet of mealworms, which they did not need to strike, responded by striking at the same stimulus objects and with the same promptness as normal subjects. There is circumstantial evidence that striking at live flies enhances the future skill of actually catching them (Rilling *et al.*, 1959). Many parasites and symbionts are guided by chemicals when searching for their host or partner

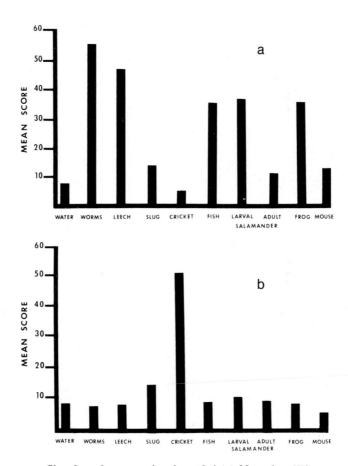

Fig. 34 a and b. Response profile of newborn previously unfed (a) 22 snakes (*Thamnophis r. radix*) to water extracts from the body surface of various small animals. (b) five smooth green snakes (*Opheodrys vernalis blanchardi*) of the same age as in (a). "Worms" and "Fish" comprise three species each. (From Burghardt, 1967. Copyright 1967 by the American Association for the Advancement of Science)

(Osche, 1963) and it is unlikely that they need to learn their meaning. Newborn snakes of many species have been shown to exhibit clearcut preferences for body surface extracts, prepared from natural rather than nonprey species, as measured by their rate of tongue flicking and genuine attack (Burghardt, 1966, 1967, 1969, 1970 a, b). Response profiles (Fig. 34) obtained from newborn specimens illustrate the food preference of two species of snake. While in *Opheodrys* the response to all prey extracts, except that from crickets, hardly surpasses that to distilled water as a control stimulus *Thamnophis* proves truly euryphagous. By contrast, Morris and Loop (1969) claimed that newborn rat snakes would respond equally to extracts from nine prey animals though significantly more strongly than to the water control stimulus. However, Campbell (1970), using a slightly different method in chemical-cue experiments with the same species obtained a clearcut preference for frog odor which is commensurate with the diet of the species at that age. Likewise, Burghardt (1973) demonstrated preferences for prey in newborn common five-lined skinks (*Eumeces fasciatus*), that captive specimens accept as food (see also Loop and Scoville, 1972). Recently, Burghardt and Abeshaheen (1971) working with *Elaphe vulpina* also demonstrated a clearcut preference for mouse, the natural food of newly hatched young, and could at the same time show that the conflicting results of Morris and Loop (1969) had most likely been due to the lack of control for habituation. In further experiments Burghardt (1970 a) was able to show that species-characteristic food preferences persisted for up to six months, during which the young were force-fed strained liver baby food with the aid of a cannula. Accordingly preferences persist in the absence of reinforcement from sensory cues associated with normal feeding.

Despite their long-term persistence against extinction, prey preferences can be experimentally modified, so as to encompass a new preference for a food that can never be encountered in nature. Newly hatched *Thamnophis butleri* were exclusively fed horse meat covered with worm extract, i.e. an odor cue preferred by the species. After six months of progressively decreasing the worm extract to zero horse meat extract still received the highest score (Burghardt, 1969). Persistence of this artificially induced preference was not examined. Though the broadening of releasing stimuli involved a procedure that cannot conceivably happen in the wild, some modification might occur since an accentuation for a certain prey preference was obtained under conditions that might well operate in nature. When newborn *Thamnophis sirtalis* were fed with one of two favorite prey (worms versus fish) and were tested thereafter, some exhibited a marked preference for the "imprinted" prey, while some readily switched to the new food (Burghardt, 1970 a). The problem remains unsolved as to whether individuals that did not switch, had received as the first prey the one which they would in any case have chosen innately. For further modification of prey preferences see Chap. 3. D. As will be seen below (Chap. 5. B. III. 2., Fig. 63) the ensuing distribution of individuals with broad or narrow response profiles cannot be distinguished from intraspecific variation of prey preferences which has come about without feeding experience.

Hence, work on chemical-cue preferences confirms that newborn snakes and skinks are equipped with an innate releasing mechanism [6] for the recognition of prey that forms their staple diet in the wild and/or in captivity. The arguments presented above and similar results are incompatible with Hinde's (1958) hypothesis of food recognition only through trial-and-error learning, though it may well be valid in some cases (e.g. Hogan, 1971).

Evidence for innate recognition of prey appears compatible with a modification of Hinde's (1958) hypothesis: while certain common characteristics, shared by many prey are recognized innately, trial-and-error learning would furnish the details necessary for recognition of individual prey species. After recognition of individual prey species had been acquired the predator might generalize, so that without much further learning the prey spectrum could become very large. This generalization hypothesis appears to be the simplest possibility in terms of innate capacities required, and is supported by two types of evidence which are admittedly indirect: (a) After an experience with a few insect and flower patterns, two primate species *(Cebus capucinus, Macaca mulatta)* were able to generalize quickly to large numbers of other insects or flowers; an innate recognition of both patterns was barely evident (Lehr, 1967; see also p. 87). (b) Large samples of insect prey of anurans and tyrant flycatchers (Tyranni) proved taxonomically highly diverse but of rather uniform body size (Berry, 1965, 1970; Hespenheide, 1971). In the case of birds, this unspecificity is certainly not due to some sensory inferiority: swallows and kingfishers in flight feed on honeybee drones but reject the noxious workers (Beal, 1918; Rothschild, 1971).

Yet there are facts that cannot be accommodated by the generalization hypothesis. For common toads with an extended and diverse experience of live prey in the wild, movement remains such a powerful attack-eliciting stimulus, that any moving object of acceptable size is attacked and rejected only if inedible. Thus common toads have to learn from a vast array of objects what not to eat (Eibl-Eibesfeldt, 1951). In this case neither generalization from previously encountered prey to future prey nor an individual innate recognition seems to hold. To some authors the alternative view, i.e. an individual learned recognition, appears untenable: "It seems inconceivable that each of the many diverse types of stick, leaf, fecal pellet, moss, lichen, and so on is recognized individually as being inedible" (Robinson, 1969 b, p. 252; see also Hinde, 1958). Yet in the case of the toad referred to, such individual learning seems the only reasonable assumption.

II. Capture-inhibiting Prey Stimuli

In general, prey is recognized by specific or generalized stimuli eliciting attack (Chap. 3. A. I.). However, a number of euryphagous predators are faced

---

[6] Perceptual mechanism, underlying the readiness to respond selectively to certain stimuli, more than to others, independent of individual experience with those stimuli.

with the problem of exempting from attack such animals as are either harmful or beneficial in terms other than food, but would appear to possess properties that provoke predatory attack. In many cases no special solution to this problem has been achieved, as is borne out by the widespread occurrence of cannibalism. But in other cases, inhibitory mechanisms designed to prevent such fatal results would seem to exist. This does, however, not imply that cannibalism, a discussion of which would go beyond the scope of this book, is universally detrimental.

Piranhas recognize their prey fish primarily by visual stimuli; mechanical and olfactory (blood) stimuli merely tend to lower the threshold for attack (though the importance of the former, now seems to have been underrated (Markl, pers. comm.)). Conspecifics would fall within the size range of fish that are readily attacked. As demonstrated by Markl (1972) by dummy experiments, piranhas tend to attack all fish shapes that exceed a ratio of length : width of 4 : 1, i.e. elongate species. They exempt, other things being equal, all fish shapes that have a length : width ratio of less than 3 : 1. In this way conspecifics with their ovoid body shape are never attacked (unless wounded or swimming abnormally). Only very few other species can be predicted to benefit from sharing the unpermitted range of length : width ratios. The functioning of such an inhibitory mechanism is particularly important in species, which like *Serrasalmus nattereri,* are permanently social. While nothing is known about the ontogeny of this simple and effective device another attack inhibition is known to be learned. Piranhas also do not attack smaller and agile species, which they know from experience to be virtually invulnerable because of their flight capacities (see also Lindroth, 1971; Hespenheide, 1973; Gibson, 1974). It is interesting that from time to time a piranha thus educated, will suddenly lunge towards an elusive fish in a surprise attack. Because of their rarity, such surprise attacks find the prey rather unprepared and thereby enhance the chance of capture (see also Steiniger, 1950: Norway rat). Thus the constancy of both, the inhibition in attacking conspecifics, and in attacking agile prey can be understood in terms of reward for the predator.

Similarly, the cleaners involved in marine cleaning symbiosis are immune from attack by their predatory hosts (review Feder, 1966). Over 45 species of fish are known to be cleaners, as well as six species of shrimp. Two factors are thought to account for the immunity of the cleaner. One is a particular behavioral or color signal emitted by the cleaner, by which the fish predator distinguishes it from genuine prey (Wickler, 1968, p. 144 seq.). The other is the unwillingness to hunt, because of satiation before visiting a cleaner (Eibl-Eibesfeldt, 1955 a) or the fact of being covered with ectoparasites (Feder, 1966). The potential host sometimes adopts special soliciting postures, e.g. above a coral rock known to harbor a pair of cleaning fish when in the "mood" for being cleaned; moreover , particular movements announce the impending departure of the host, which prewarns the cleaner to leave, e.g., the mouth (see discussion in Hobson, 1971; Potts, 1973 a).

The nature of this immunity is indicated by a revealing observation (Hediger, 1968). A large grouper (*Epinephelus* sp.), almost four feet long, had

94

been kept in a tank from infancy for six years. The grouper was accustomed to snap up any food that was thrown into its tank. One day a live cleaning wrasse (*Labroides dimidiatus*) was dropped into the tank instead. The grouper not only failed to snap up the cleaner but opened its mouth, and allowed the cleaner free entry and exit, in spite of the absence of ectoparasites. It even tolerated two more cleaning wrasses and eventually fled from their unnecessary "cleaning" efforts. These observations dismiss any hypothesis that the cleaner's survival depends on host satiation or parasitization. There are species that are "agressive mimics" of the cleaner and rely on the passive behavior of fish, which suppose they are about to be cleaned, to dart in and bite off a chunk of their fins (e.g. *Aspidontus taeniatus*). The existence of these mimics, which resemble their models in appearance and initial approach behavior, is evidence of strong natural selection for hosts with no intention of harming their cleaners (Wickler, 1968; see also Potts, 1973 b).

Another case of rejection, on sight alone, is that of the palatable caterpillar of a neotropical hawkmoth, by an anole (*Anolis lineatopus*); it remains tantalizingly obscure, as it defies any explanation by established principles of predator avoidance. Although it moves, and is discovered by the anole, the brown larval variant is rejected, while under the same conditions, the green and the blue variants of the species are accepted. The rejection of the brown larva seems to be associated with a minimal body size. The brown larva is neither a mimic, nor distasteful, nor is it overlooked because of its beautiful twig camouflage (Curio, 1970 b). We are presently investigating the possibility that the larva is protected by a kind of camouflage, that operates *despite* the movement of the caterpillar, which would lead it to be detected by the anole.

In conclusion, it would appear that predators utilize both eliciting and inhibitory stimuli from prey animals, to recognize them properly. A situation such as this closely parallels food plant recognition by euryphagous herbivorous insects, which respond by orientation prior to feeding, either to an optimal combination of nutrients *or* to the absence of repellents (review Lindstedt, 1971).

B. One-prey : One-response Relationships

Predators assume different hunting tactics with different prey and/or under different circumstances. A predator is able to discriminate between at least as many prey animals as its number of prey-specific hunting tactics suggests. In a most extraordinary way some predators show prey discrimination, in that they undergo morphological changes induced by and adapted to different prey. For example, the same individual protozoon, *Glaucoma vorax*, assumes a slender form when feeding on bacteria or yeast but becomes a giant in the presence of larger prey, i.e. other live ciliates (Fig. 35), or becomes a cannibal when food is short. While giant formation upon ingestion of larger prey is not uncommon among protozoa (see also Giese and Alden, 1938; Njine, 1972) *Glaucoma vorax* anticipates ingestion and digestion of the larger prey in response to chemical or mechanical stimulation from it; it grows enormously and forms a preparatory vacuole. Giants can

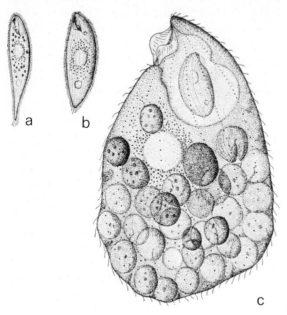

Fig. 35 a – c. Modifications of the ciliate *Glaucoma vorax* as a result of the perception of different prey organisms. ×500. (a) tailed form, bacteria-feeder; (b) saprozoic form, feeding e.g. on yeast; (c) large carnivore feeding on other ciliates (*Colpidium, Glaucoma, Colpoda*). Note the preparatory vacuole near the enlarged mouth. (After Kidder *et al.*, 1940, courtesy of Biological Bulletin, Woods Hole/Mass.)

revert by several divisions to smaller forms (Kidder *et al.*, 1940). It is remarkable that comparable dietary changes, involving a similar morphogenetic flexibility, occur in Fungi preparatory to trapping nematodes (Pramer, 1964), in Rotatoria (Gilbert, 1973), tyroglyphid mites (Woodring, 1969), and even Amphibia (see Chap. 5. B. III.), when becoming cannibals.

C. The Assessment of the Circumstances of a Hunt

Apart from the type of prey, the circumstances of a hunt affect hunting success, and need therefore be correctly assessed by the predator. Constraints taken into account by predators range from the obvious physical ones to the prey's prospects of escape. Responses of the predator involve an adjustment of hunting method (see Chap. 5. B. II.), or simply affect the rate of capture attempts. For example, for birds securing their fish prey by plunge-diving, water surface conditions are of considerable importance. Brown pelicans (*Pelecanus occidentalis*; Robinson, pers. comm.) and terns (*Sterna hirundo, S. sandvicensis*) (Dunn, 1973 b) tend to fish more when the sea surface is rough (not stormy). A higher wind speed and a rougher surface independently lead in terns to a higher rate of diving and fish capture than a lower wind speed and a calmer surface. It is important to note that in general, a rough surface leads to an increase of hunting activity, and not just of capture success. Dunn (1973 b) thinks that a rough surface impairs a fish's view of a hovering tern more than the tern's view of the fish, so that dives result in more successful captures. He bases his assumption on the experience of divers, who recount "that it is easier to see someone below the surface from above, than it is for the person below the surface to see someone above the surface". Furthermore, with more wind, terns need not hover so vigorously and consequently will not be detected as easily by their prey fish. Yet any generalization would be premature; king-fishers stop fishing when it begins to rain and the water surface becomes rippled, and perhaps less penetrable to sight (Kniprath, 1969), though rain may be detrimental to fishing, in other ways too.

The perceptual performance involved in the assessment of the capture success makes it fully clear that talking about "prey recognition", in terms of prey characters only is an oversimplification. An evaluation of the often highly complex circumstances of the hunt to follow must also be involved.

D. Prey Recognition by Prey-related Signals

Closely related to the last-mentioned performance of predators is recognition of prey by signals associated with it. As a rule, such signals permit a crude localization, so that the predator can concentrate its subsequent search onto a restricted area. In alerting a hungry predator the far-distance senses also play a prominent role.

Cues Emanating from the Prey or Its Surrounding. Often odor from a prey betrays its presence to a searching predator. Tracking down a prey, involves either the straight approach toward the source of odor or "working up" a track which the prey has laid down. Muricid snails (*Urosalpinx cinerea, Eupleura caudata*) are attracted over several hundred feet by fields of oysters (*Ostrea sp.*), and the accurate localization of the individual oyster is also steered chemotactically (Galtsoff *et al.*, 1937; Carriker and van Zandt, 1972; see also Zafiriou *et al.*, 1972). Olfactory discrimination may occasionally exceed even visual recognition in precision: *Urosalpinx cinerea* can distinguish between prey which differ only by previous nutrition; it prefers barnacles that have fed on *Artemia* larvae to others that have fed on algae (Wood, 1965).

Many predators follow odor trails left behind by their prey. They would expend much less energy if they could tell from a trail which direction the prey has taken. Three species of snails, a prosobranch (*Littorina irrorata*) (Hall, 1973), an opisthobranch (*Nassarius obsoletus*), and a pulmonate (*Physa acuta*), discriminate between the two directions of an odor trail, in these two latter species regardless of whether they follow their own trails or those of conspecifics (Crisp, 1969; Wells and Buckley, 1972). Following a trail in the direction of the trail-maker enables snails to move quicker, so that trail-following may be a matter of economy in locomotion (Hall, 1973). In no case is it known, whether this capacity is used to track prey which marine opisthobranchs in fact discover by olfaction and by following odor trails (Paris, 1960; Paine, 1963; see also discussion Wells and Buckley, 1972). The European asp (*Vipera aspis*) is known to sense by Jacobson's organ the odor trail of an envenomed mouse after having struck it with its poisonous fangs. The possibility that a viper recognizes the direction of the odor trail can be virtually ruled out (Naulleau; 1966; Burghardt, 1970 a) but remains to be investigated in other snakes (see Watkins *et al.*, 1967). There would appear little need for such a performance, as the length of the victim's trail is relatively short and nonconvolute.

Hearing is another prominent far-distance sense employed in finding prey. Sharks follow unusual or escape movements of prey animals from up to 200 m away by virtue of their lateral line organ; when close to a potential victim, they may use a number of senses in combination before actually attacking. It is still open to question whether the lateral line organ can also be used alone, to localize a prey accurately (Banner, 1972; Nelson and Gruber, 1963).

As a rule, prey animals try to hide their presence by adaptive silence. Fear of a nearby predator usually suppresses vocalizations still more effectively, unless the prey is caught and emits distress calls. These in turn may attract other predators. Thus birds of prey (Diezel in Sunkel, 1927; Peeters, 1963) and crows in large numbers (Diezel in Sunkel, 1927; Kramer, 1941) are attracted upon hearing the distress calls of a wounded hare (see also Busnel, 1963, p. 90 seq.). It is unknown whether it is actually rewarding for a predator to approach distress calls; as they are called forth from a quarry already caught by a competitor, any predator arriving later will probably go unrewarded. Only if the first predator relinquishes, can a meal be obtained (see p. 160).

Visual cues betraying the presence of a prey may be extremely subtle and perhaps only employed after prolonged training. For example, herring gulls specialize, in detecting shore crabs buried under sand when the water is falling. During low tide these crabs are covered by barely visible flat "domes" almost level with the surrounding beach. As demonstrated experimentally, the specialist gulls discover many of the crabs by looking for these "domes" (Shaffer, 1971). It would also be interesting to know whether predators are able to reject structures that appear potentially rewarding but which show subtle signs of depletion from an earlier attack (e.g. see Fig. 29). As reported previously, certain insect parasitoids avoid attacking host pupae that a competitor has already attacked (p. 43). The utilization of indirect prey-related cues by the predator in tracking down the prey is at variance with learning experiments under artifical conditions; patterns that were spatially separated from the reward, if only slightly, could only be conditioned after long periods or not at all. This applies to such diverse animals as lizards *Lacerta* sp. (Svoboda, pers. comm.), birds (Schuler, 1974), and primates (review Lehr, 1973). The discrepancy could not be reconciled by assuming that responses to prey-related cues in nature need no conditioning: the herring gulls must learn what crab "domes" look like. But it could be reconciled by assuming that in nature, the learning of prey-related cues involves a predisposition to learn certain things much more rapidly than others. This would imply that the laboratory situations mentioned have been too artificial to produce the performance intended by the experimentor.

Cues Emanating from Nonprey Animals. Animals that form feeding aggregations provide important signals that facilitate prey detection. Hearing and vision seem to play a prominent role. Thus, Frings *et al.* (1955) found that herring gulls use a food-finding call, which lures conspecifics from as far as 5 km away across open water. A typical flight-pattern of gulls that have found food, or the formation of an "interest-group" around food, provide additional visual signals. Gulls first lured by one or the other signal then locate food directly by vision. In a similar way piranhas (*S. nattereri*) launch an attack in a direction in which they see other school members darting forward, even though they have not seen the attack-provoking stimulus themselves (Markl, 1972). Such a social facilitation has been also suggested for the communal attacks of squid and chasing fish predators (refs. in Neill and Cullen, 1974). It benefits the social community only if (1) communal hunting is more successful than individual hunting, as is true for some predators (see below Chap. 5. C. III. 3.). (2) When food is limited the cost paid for attracting other individuals to a source of food, and thus aggravating competition, must be compensated for by being equally often attracted. To account for such "reciprocal altruism" by individual selection is sufficient (Trivers, 1971). While condition (1) may be a possibility to facilitate social attack in piranhas, condition (2) is likely to apply to the herring gulls mentioned earlier; they are long-lived enough and live close enough to each other to enjoy the benefits of "reciprocal altruism"; moreover, condition (1) can be dismissed for herring gulls, since their feeble prey items do not require communal attack.

The attraction of other foragers is often not in the interest of those providing the betraying cues. This may be so, because of increasing food competition when "reciprocal altruism" is of no avail, but also because of threats produced to close relatives. An example of the first possibility is found in lions and hooded vultures (*Necrosyrtes monachus*), which follow the screaming and giggling of spotted hyenas around a kill, and when close, lions eventually appropriate it. Since the chances for appropriation are not equally reciprocal, to attract other foragers seems to be disadvantageous (Kruuk, 1972 b, pp. 135, 146, see also Schaller, 1972, p. 213). The other disadvantage arises as a side-effect of parental care in birds. Skutch (1949), and later Snow (1970) and Brosset (1971), have advocated the idea, as yet unproven, that tropical altricial birds have smaller broods than their temperate zone relatives, because arboreal predators would locate the nest, by observing parents carrying food to it. Clearly, in this view, a reduced brood size would confer an advantage by decreasing the rate of feeding visits.

A rather indirect hint in support of this suggestion was offered by Riehm (1970), who observed long-tailed tits (*Aegithalos caudatus*) hovering in front of their nest, as if to catch insects. The mock capture is thought to distract nearby predators to observe feeding visits, and mistake them for real collection of food. This interpretation, however, would demand considerable powers of abstraction in a potential nest marauder, and other interpretations appear as likely (Gaston, 1973).

As described by Löhrl (1950 b) European jays seem to locate nests of the chaffinch (*Fringilla coelebs*) and spotted flycatcher (*Muscicapa striata*) by tracking the intensity gradient of mobbing by the parent birds. Details of Löhrl's account leave no doubt that the female European cuckoo (*Cuculus canorus*) also finds nests of meadow pipits (*Anthus pratensis*) and perhaps even uses the overall intensity of mobbing to assess the proper breeding stage of the host. As is well known, the intensity of mobbing increases with decreasing distance from the nest, so that the two interlopers may draw conclusions as to its location. Experimental proof of this highly interesting possibility would be worth every experimental effort, since it would seem to constitute the first case of acoustical orientation by kinesis.

E. Prey Stimulus Summation

Evidence relevant to the question of heterogeneous (stimulus) summation (Seitz, 1940) in prey recognition is scanty. Drees (1952) found that in a salticid spider, prey size, shape, solidity, movement, and contrast to the background all combine to stimulate capture. Omission of a stimulus parameter only weakened the response, but did not abolish it. This was also the case with prey discovery by rufous-naped tamarins, where aspects of general insect body shape and appendages were important (Robinson, 1970; see also Schuler, 1974). Similarly, the little blue heron (*Butorides sundevalli*) attacks sally light-foot crabs (*Grapsus grapsus*) living on the rocky shores of Galapagos. The heron begins to stalk when only a crab's legs become visible from behind a rock (pers. obs.).

Super-normal stimuli that elicit capture have been found in a stimulus analysis with praying mantids. Fly dummies with waxpaper wings on a paper disc were more effective than wings removed from flies, including dead "moving" flies. Moreover, when the number of "wings" on the paper disc was varied from 0 to 8, the stimulus value increased almost linearly with the number of "wings". An increase in "wing" number beyond 8 was not accompanied by a further increase in stimulus value (Rilling *et al.*, 1959). For obvious reasons, prey animals evolve, so as to stay as far away as possible, from the range of the super normal.

In few cases has the question been answered, how sign stimuli of a prey are evaluated quantitatively (Schuler, 1974). Starlings evaluate a combination of cues, of an acquired visual insect pattern, i.e. of an unpalatable Batesian model, in a non-additive manner as measured by discrimination from the model pattern (Fig. 36).

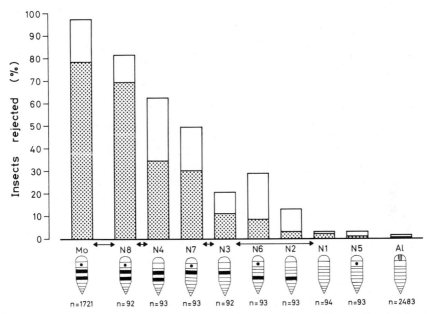

Fig. 36. Percent rejection of palatable mealworm pupae by starlings after having learnt unpalatability of model pupae *Mo*. Pupa patterns are blue except for *Al* that is brown. *Stippled:* rejection; *empty:* pecked but not eaten. $n =$ number of pupae of each sort. *Arrows* indicate statistical significance of difference of rejection between the pupae concerned. (From Schuler, 1974)

Interaction between pattern elements, as well as the absence of any one element from the mimic, merely impairs recognition but does not lead to its total breakdown (see above). Essentially this mode of stimulus interaction has also been found in avian enemy recognition by birds (Curio, 1969), thus supporting the idea that a purely additive stimulus co-action is an exception rather than the rule.

In other cases the stimulus complex offered by a prey cannot be broken down into single aspects of the whole. For example, trigger fish (*Balistes fuscus*) prey upon various species of sea urchins, by blowing a jet of water sideways onto their body, so that they can bite into the vulnerable oral disc, break off the spines, or lift the whole animal by them. Wooden dummies with wire spines were effective in eliciting the attack, only if consisting of both stimulus parameters, i.e. sphere plus spines. This applied to both an intact dummy, when offered singly, as well as when offered together with a defective one, devoid of spines or totally lacking the central body (Fig. 37; Table 3).

101

Fig. 37. A trigger fish lifts an intact sea urchin dummy from the sea bed while neglecting a defective dummy devoid of spines. (From Fricke, 1971)

Table 3. Results of sea-urchin dummy experiments, carried out with three adult trigger fish. Response scored by fish blowing sideways at the dummy in order to lift it and thus expose the vulnerable oral disc. (x) number of trials. For further explanations see the text. (From Fricke, 1971)

Dummy	Response
☀	9 (11)
☀	0 (16)
●	0 (12)
☀ : ☀	7 : 0 (7)
☀ : ●	0 : 0 (7)
☀ : ●	9 : 0 (9)

☀ Sea urchin dummy intact
● ″ ″ spineless
☀ ″ ″ all-spines ″

This result applies to the stationary dummy. However, a sea urchin when buried in the sand becomes immediately attractive when it moves the spines protruding above the surface; if this is not the case, it is ignored, as is the bodiless wire dummy (Fricke, 1971). If such a dummy is to be fully comparable to the buried live sea urchin, it follows that spines plus body are indispensable when motionless, while the body is no longer necessary once the spines begin to move.

It would be worthwile to record the conditions, under which prey is recognized by its parts or by the whole *Gestalt*. It is tempting to speculate that prey moving quickly, as for example the sally lightfoot crab mentioned, or one that is commonly concealed from view, is already recognized by its parts, a perceptual performance that would be clearly adaptive.

Strangely, no summation of prey-stimulus parameters has been found in hand-raised loggerhead shrikes. The birds responded to a rectangular wooden peg on the ground at an age when mouse-killing commenced. Vigorous pecks were delivered to an end of the peg, that had a bilateral indentation ("neck"), or had two symmetrically arranged black "eye" surrogates. A peg with identical ends received equal numbers of pecks at each end. When such a peg was moved along its long axis, pecks were delivered to the front end. When movement and pattern were added in another trial, the response, as measured by the percentage of pecks to the preferred end, did not increase beyond the response obtained by movement or pattern alone. This is especially puzzling, in view of the fact that neither aspect had yielded the maximum response (S. Smith, 1973).

F. Novelty Versus Familiarity

The only rule that would thus far predict the acceptance or rejection of a novel food, by an animal, is an extremely general one: As experience with novel food continues to grow, novelty gives way to familiarity, leading eventually to ingestion.

I. The Rejection of Novel Prey

Potentially edible prey animals, on the first encounter, are rejected by predators, such as orb web-spiders (*Achaearanea tepidariorum*) (Turnbull, 1964), fish (Ivlev, 1961; Springer and Smith-Vaniz, 1972), reptiles (O. v. Frisch, 1962), birds (Swynnerton, 1919; Hogan, 1965; Coppinger, 1969, 1970; Morrell and Turner, 1970), and mammalian carnivores (refs. in Gossow, 1970; Ewer and Wemmer, 1974). One of the causes of primary rejection may be the arousal of fear. Domestic chicks (Hogan, 1965), a number of songbirds (Coppinger, 1970), owls (Herrlinger 1973), and mustelids (Gossow, 1970) responded with signs of fear, e.g. uttered alarm calls or even panicked, on seeing a novel prey. The phase of rejection, if it occurs, can therefore be understood, on the assumption that the sight of a novel prey arouses attack readiness (central prey excitation) and at the same time fear, and that initially, fear would dominate. Subsequently continual (or repeated) stimulation from the prey would lead to the habituation of fear responses, so that attack responses could predominate. On a modified view, a novel stimulus engenders "general arousal" (Berlyne, 1960), and subsequent habituation to the stimulus would reduce arousal to levels permissive for attack (see also Hogan, 1965, 1966).

Modifying an enumeration by Coppinger (1969), one can discern four mechanisms that can account for a non-random selection of food: (1) Selection of familiar food, or, similarly, an innate avoidance of novel stimuli (see Chap. 2. C. III.); (2) Innate preference for specific sign stimuli (p. 90 seq.); (3) Innate avoidance of specific sign stimuli which cannot be overcome by familiarization; (4) Generalization from previous noxious experience, e.g. from unpalatable prey. Note that (1) leaves open whether novelty is judged by the animal in relative or absolute terms.

In order to find out the reasons for rejection of novel prey (1, 3), Coppinger (1970) hand-raised 30 insectivorous birds (17 blue jays (*Cyanocitta cristata*), 7 common grackles (*Quiscalus quiscula*), 6 red-winged blackbirds). These were tested at an age of 9 – 10 months after having received two sorts of pelleted food, but no insects. Test insects, mealworms at first and then three species of neotropical butterflies of comparable palatability, were all offered in a predetermined sequence. At first the birds rejected the mealworms to varying degrees, and on the average, took longer than wild-caught conspecifics to accept them finally. Subsequently, half the birds were presented with 18 of one butterfly species, and thereafter 18 of another, while the other half received the same insects in the reverse order (Fig. 38); while *Anartia jatrophae* is white and brown, *A. amalthea* is black and red. Although both butterflies were equally new to the birds, *A. jatrophae* was accepted significantly more often than *A. amalthea*. Moreover, the experience with the first species, influenced the acceptance of the second in different ways. As suggested by the fraction of birds accepting at least one butterfly in both trials, a high rate of

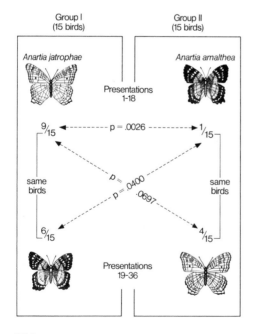

Fig. 38. Comparative acceptance of two species of butterflies related to the order in which they were presented to butterfly-naive individuals of three species of insect-eating birds. Figures denote number of birds accepting at least one of the butterflies. For further details see the text. (Modified from Coppinger, 1970, by courtesy of Am. Soc. Naturalists)

acceptance of the first species raises that for the second one, and vice versa. One must consider, however, that the two groups of birds are not strictly comparable, because of experiential differences prior to testing (Coppinger, pers. comm., 1973), and because six birds of group I which ate *A. amalthea* were amongst the nine that had eaten *A. jatrophae* previously; by contrast, most of the birds of group II which ate *A. jatrophae* had had only visual experience with the highly rejected *A. amalthea*.

In a final experiment seven birds from both groups that were freely feeding on both butterfly species were presented with a third and larger black, orange, and yellow species (*Protogonius hippona*). Only two of these attacked the large butterfly on first presentation while the rest refused it or accepted it only later.

These results were taken by Coppinger (1970) to mean that previous feeding experience shapes rejection or acceptance in the birds' subsequent life. For before the experiment the birds had been fed brown mealworms and brown pellet food and thus were believed to have generalized to the white and brown *A. jatrophae* and not to the very different *A. amalthea*. (A version of mechanism (1), see above). The refusal of the lastly offered Protogonius was explained by a lack of generalization from the first two butterfly species in size, shape, color, and pattern. Rejection or acceptance of a novel prey was thought to be determined by the "amount of stimulus change and the past experience of the animal, and are not related to any particular characteristics of the stimulus per se" (p. 331), thereby invoking mechanism (1).

The results of Coppinger are, however, open to at least one other interpretation. They are based on the ad hoc hypothesis that the birds would judge on similarity in terms of color rather than in terms of size and shape, but the contrary might be true. All three butterflies might actually be regarded as more alike than the first two butterflies and the previous food (mealworms, pellets). Moreover, a crucial control experiment was omitted, i.e. there were no birds that had been pre-fed a training food resembling the highly rejected *Anartia amalthea* in coloration, which would have been a necessary condition to establish the conclusions drawn. The results obtained by Coppinger could be equally well explained by assuming an innate inhibitory mechanism responding more strongly to *A. amalthea* and *P. hippona* than to *A. jatrophae* (3). This explanation would leave room for generalization to occur but would not make it of crucial importance. Moreover, such an interpretation would seem compatible with the fact reported by Coppinger (1969) that wild-caught blue jays learned to accept *A. amalthea* more slowly than *A. jatrophae* though both novel and training food were the same in both cases. It should be noted, however, that different training diets did have different effects on the reactions of these jays to new insects. Therefore any theory that attempts to account for the responses of birds to novel prey has to consider past experience with familiar prey.

Rejection of novel and even of aposematically colored prey is by no means universal, as shown by freshwater fish (*Gambusia affinis, Carassius auratus*); they had to "taste" highly unpalatable, marine food organisms of such coloration (*Pleurobranchus*) before rejecting them (Russell 1966).

Rubinoff and Kropach (1970) made a pertinent study on the avoidance by predatory marine fish of the brightly colored and poisonous sea snake (*Pelamis platurus*), found only in the Pacific Ocean. Some Pacific sympatric predatory fish consistently ignored the sea snake, but captive fish from the Atlantic readily attacked the snakes and some even died as a result of bites. Rubinoff and Kropach inferred that selection had favored Pacific fish with an innate reluctance to prey on this species (3).

Apart from the question of which mechanism exactly governs the rejection of novel prey, these studies lead to a reconsideration of the theory of warning coloration, in that predators need not always become conditioned to warning colors in order to avoid them later on. Rather, rejection may be based upon an innate recognition of certain prey aspects, although the available evidence for this notion is not very impressive.

Finally, Rabinowitch (1968) showed that ring-billed gulls prefer a familiar diet to a novel one and that from two novel diets a bright green food was never selected while a pink food was. These results bear out that an innate inhibitory mechanism, that responds to particular stimuli (3) and one that responds to novel stimuli (1) may occur side by side in the same animal, together with an innate preference (2). It remains to be investigated how these mechanisms relate to each other and to generalization of obnoxious experience with prey (4).

II. Familiarization with Prey and Its Consequences

Prey that is initially rejected may be accepted later because the predator has become familiar with it. Familiarization need not involve physical contact with the prey. After a "latent phase" during which novel prey is rejected on sight alone it will suddenly be eaten. In captive rainbow trout the "latent phase" was found to last, on average, four days (Ware, 1971, 1972; see also Beukema, 1968) but in songbirds it may take only a few seconds (e.g. Morrell and Turner, 1970). Still other predators do not noticeably hesitate before taking the first of a given type of prey. The amount of time that the prey is exposed during the "latent phase" may be influential in terminating it. The possible importance of this factor is supported by but very few observations. Given a permanent view of their prey, captive pigmy owls are, other things being equal, less ready to hunt than with intermittent exposure (Scherzinger, 1970, p. 29). A more detailed investigation of the problem is badly needed.

In some cases an accentuation of innate prey preferences is based upon feeding responses (e.g. p. 92). In an attempt to separate the rôle of food reinforcement from that of mere exposure to the relevant chemical stimuli eliciting feeding responses in young garter snakes Burghardt (1970 a) found performance of capture and ingestion to be of paramount importance. Mere pre-exposure to the relevant chemical stimuli alone did not accentuate preferences as evidenced by individuals whose prey-attack did not habituate by stimulation per se.

Familiarization gives precedence to one type of prey over others that have not yet been encountered by the predator. The question arises as to how far this effect is permanent. In the extreme, the primary food will be accepted to the exclusion of any other encountered subsequently. Some young garter snakes (Burghardt, 1970 a; Fuchs and Burghardt, 1971) and some young domestic chicks (Rabinowitch, 1965) of experimental groups that had each been reared on one type of food did not switch and starved to death. Cases such as these would fulfil one of the basic criteria for imprinting, i.e. that of irreversibility (e.g. Hess, 1959, 1964; Immelmann, 1972). It must be borne in mind, however, that in nature results like these would hardly occur because most euryphagous predators have access to a much more varied diet. Evidence for less fatal effects of early experience, resulting in mere prey *preferences* have also been obtained in other species, namely *Acanthocyclops vernalis* (Brandl, 1973), herring larvae (Rosenthal, 1969 a), snapping turtles (*Chelydra serpentina*; Burghardt and Hess, 1966), red-eared turtles (*Pseudemys scripta*; Mahmoud and Lavenda, 1969), owls (Scherzinger, 1970, p. 30), and mustelids (Gossow, 1970; Apfelbach, 1973). The lability of such preferences is highlighted by prolonged experimentation by Mayer and Quednau (1958, 1959). By raising *Trichogramma* sp. for many generations on a particular host species, they could drastically change the preference for the host which normally is *Galleria*, *Cimex*, or *Arctia*. The least preferred host (*Galleria*) could, after 53 generations of breeding on its eggs, be made the most preferred with a wholesale acceptance of its eggs. This change of preference ("training bias") is obviously nongenetic since after a single passage through *Cimex* in F_{111} seven from 20 females again refused to lay into *Galleria*. It remains to be explored which factors accumulate during many generations of enforced breeding that change a preference so resistant to modification. A similar "training bias" has been reported for Japanese oyster drills (*Ocinebra japonica*) (Chew, 1960) and, mainly, vertebrates in captivity (see Chap. 2. C. III.). Reports on more stable preferences are noticeably scarce but there are indications of prey imprinting taking place. Significantly, adult water snakes (*Natrix sipedon*) seem to be less affected by recent feeding experience than young ones (Gove, 1971 in Burghardt, 1975). This finding might indicate the existence of what is called the "sensitive" or "critical" period of true imprinting phenomena. Working with odors from three different types of prey Apfelbach (pers. comm.) found that young polecats exhibit a preference for the odor of a prey on which they had been exclusively fed. In addition there was a common basic food of chicks for all experimental groups. Preferences were the more marked the longer individuals had fed upon the prey in question. Further work must discriminate whether this is due to the length of feeding experience or to age. From his experiments Apfelbach concluded that any prey odor must be learned by prey-naive polecats. However, Gossow (1970) found in studies on both polecats and stoats (*Mustela erminea*) that prey-naive young animals, while uttering social contact calls, as in pursuit of litter mates, pursued potential prey animals. When reaching the prey they seized it as they would have seized a litter mate during play. But upon contact with the prey they immediately switched to typical killing behavior, i.e.

strived at a position to apply the killing bite. These observations suggest that young mustelids respond, appropriately, on first encounter, to (odor plus tactile) stimuli from the prey. The failure of Apfelbach's animals to do so would appear to be due to his scoring mere search behavior, that can indeed be elicited by odor alone.

The suddenness with which the range of effective stimuli changes in some cases is reminiscent of learning during a "critical" period and hence of imprinting. Observations of young eagle owls (*Bubo bubo*), summarized by Herlinger (1973), suggest that live prey needs to be caught between 80 days and about 150 days of age for prey capture to function properly. If, up to this latter age, merely dead prey has been consumed later on the owl will capture live prey only with the greatest difficulty, or not at all. What is called "critical period" in this context may concern not only the perceptual mechanisms of prey recognition but the performance of the capture act as well, or predominantly so, as reported above for the accentuation of garter snakes.

Modification of innate perceptual capacities may take either of two forms: The range of attack-eliciting stimulus situations is broadened or it is narrowed down as a consequence of learning particular aspects of the situation. For example, full prey capture in young tawny owls will be elicited by a prey dummy, but only until the first capture of real prey (Meyer-Holzapfel, 1953). Obviously, in the process, details of the prey are acquired which render any dummy subsequently insufficient. By contrast, an adolescent polecat recognizes by sight its first rat, but only if it moves, and after having actually caught one, will attack one that does not move (Eibl-Eibesfeldt, 1958 a).

Feeding experiments with sand lizards (*Lacerta agilis*) by Svoboda (1969) have shed light upon how various stimulus parameters of the prey are processed as a function of the presence of others. When feeding for the first time, only visual stimuli of mealworms elicit attack, while the respective odor alone fails. Later on it may well become a relevant signal, which even in the absence of visual prey stimuli triggers snapping and food searching.

Among visual stimuli of a simple figure behind a bait, conditioning to color cues always occurs whereas form cues of the figure become effective only if color is absent, and even then only to some extent. Similarly, odor becomes a CS (conditioned stimulus) only if visual stimuli are absent, i.e. when the prey item had been rendered invisible or when subjects had been blindfolded (Table 4). Learning performance in the latter situation does not attain its peak with an artificial odor, whereas it does with a positive response to mealworm odor. A further result concerned the interaction of cues. Form and odor if offered in combination cannot be separated in subsequent tests on learning performance.

Quite different results were obtained with live, unpalatable insects. These were rejected after some 10 to 20 experiences. The final rejection of these prey items is based upon an evaluation of color and form, or on form alone if color had been experimentally eliminated. Further, odor as an aversive signal becomes similarly ineffective as in presentations of a dead bait. Hence the recognition of live insects as compared to simple food-associated figures is based upon different stimuli and upon a different interaction of stimuli

Table 4. The utilization of prey stimuli offered in combination with others by sand lizards after learning. + Snapping at, – or avoidance of stimulus combination; (–) avoidance incomplete. Apart from live insect experiments odor was either artificial (rose oil) or natural (mealworm juice). Live insects were unpalatable *Coccinella septem-punctata* and *Pyrrhocoris apterus;* baits were pieces of mealworm. (After Svoboda, 1969)

	Conditioned stimulus if combination was			
Prey object	color form	color form odor	form odor	odor
Bait upon simple geometric figure or odor alone	color –	color –	form odor (–)	odor + (–)
Live insect	not tested	color form –	form –	not tested

responded to. This cautions against any experiment conducted with artificial or too simple food stimuli that aims at elucidating the role of stimulus selection and stimulus interaction. How stimulus summation (Chap. 3. E.) changes as a consequence of experience with prey, remains to be largely explored.

Murdoch (1971) and Murdoch and Marks (1973) have examined the complex relationship of switching to another prey as a function of prey preference and prey density. Switching of predators means reversibility of a food habit. The existence of switching, together with a critical period possibly being rare (see above), would suggest that prey preferences only rarely result from imprinting. An euryphagous predator would hardly profit if primacy ruled over recency: Such an organization would virtually prevent the predator from exploiting the whole spectrum of potential prey species. On the other hand, weak or temporary prey preferences might well help to lessen intraspecific competition.

G. The Multi-channel Hypothesis of Prey Recognition

If different prey elicit different capture tactics (Chap. 3. C.), one must postulate different afferent channels that are tuned to different types of prey. If different prey elicit the same capture response, the question of distinct afferent channels is not easily settled. If, as can be readily shown, different prey are recognized by different sensory modalities, one can be certain that there are at least as many afferent channels as senses. In this case, the inference rests upon the structural and functional diversity of the sense organs involved. But what happens in the common case in which diverse prey elicit one and the same feeding response by one and the same sensory modality? This situation is reminiscent of the visual recognition of a number of diverse predators by distinct though homomodal IRMs (Innate Releasing Mechanisms), tuned to as many types of predator (Curio, 1975). A "multi-channel hypothesis" of this kind could be envisaged also in the present context (Fig. 39). Either two different prey animals A and B enter one general IRM (hypothesis I) or they are evaluated by two or more IRMs each of which is coping with the prey for

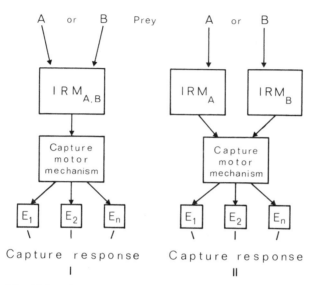

Fig. 39 I and II. Diagrammatic representation of two hypothetical relationships (I) and (II) between the decoding of different prey *A* and *B* and the motor command unit for prey capture. See text

which it has been designed (hypothesis II). The afferent signals would have to converge, at the latest, at the common capture motor mechanism, that in turn co-ordinates n effectors involved in the capture response. Ethological evidence that allows decision between the two hypotheses I and II is necessarily indirect.

a) One piece of evidence is the distribution of preferences for different prey across species of snakes (Fig. 40). Two congeneric species of *Thamnophis* exhibit different preferences under identical conditions; *T. butleri* prefers both fish and worms, while *T. elegans* recognizes fish as the only prey. This species difference is paralleled by food preferences shown in the field and in captivity (Burghardt, 1967). Thus there is selective exemption by *T. elegans* of a type of prey that is avidly eaten by a close relative, and this exemption does not impair recognition of the only prey accepted by this snake. Because of the intimate interaction of sign stimuli within most perceptual mechanisms studied so far, the selective failure to recognize a particular type of prey would be best explained by the assumption that functionally independent channels transmit information about either type of prey (hypothesis II). Such an organization would seem to permit a selective drop-out or selective formation during evolution of perceptual capacities that are tuned to given classes of prey; existing perceptual capacities of vital importance would be shielded off from such changes. A similar explanation would apply to prey preferences and differences between populations of *T. sirtalis* (Burghardt, 1970 b) and between species (Fig. 40) since the respective response profiles differ qualitatively from each other.

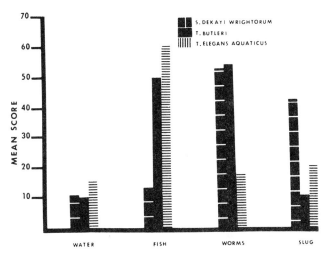

Fig. 40. Response profiles of newborn, previously unfed snakes tested with body surface extracts of three classes of potential prey. Response score derived from tongue flicks and response latency. Prey classes "fish" and "worms" involve three species each. (After Burghardt, 1967. Copyright 1967 by the American Association for the Advancement of Science)

b) Individual newborn garter snakes from the same litter preferred either fish or worm surface extracts. This variation in preference provides particularly strong support for hypothesis II. While response scores of individual snakes to two different worm or two different fish extracts were highly correlated, responsivities between worms and fish were *not*. This finding of Burghardt (1975) highlights the postulated high degree of functional independence between prey-specific perceptual channels.

While the mother's food during gestation had no influence upon prey preferences of the young (Burghardt, 1970 a), prolonged feeding of worms to newborn garter snakes enhances, on the average, their response to worms and diminishes that to fish and vice versa. This selective accentuation of one type of prey supports the multi-channel hypothesis as it bears witness of the functional independence of the channels envisioned. Further results pose problems, however, that are not easily reconciled on the assumption that channels are unitary. Accentuation obtained by monotonous feeding of worms versus fish proved to be specific for the prey used and not for the class as a whole. Thus young that had been fed nightcrawler (*Lumbricus terrestris*) did not generalize to redworm (*Eisenia foetida*), and subjects that had been fed guppy did not generalize to minnow (*Notropis atherinoides*). Thus all prey that had not been used as food scored markedly lower in individuals fed with it (Burghardt, 1970 a). Accordingly, learning processes seem to lead to differentiation within each envisaged channel. Apart from its relevance to the present problem this result lends support to the assumption that upon feeding, food preferences become modified to incorporate information about individual prey items with no recognizable generalization from one to the other. It remains to be seen whether stimulus-specifity of this magnitude is the exclusive domain of chemical-cue preferences. While for visual multi-channel mechanisms distinction of the channels seems to be of a central nature, peripheral distinction of channels appears at least possible in olfaction, in the present case at the level of Jacobson's organ.

c) Stimuli evaluated by a RM (Releasing Mechanism) would be expected to follow the law of heterogeneous summation in its qualitative form. Accordingly stimuli shared by different prey and known to elicit capture should operate identically, for example, enhance the releasing value of each kind of prey. If this rationale is accepted, then the differential evaluation of the same stimulus in different prey hints at more than one RM. As an example, tawny owls and long-eared owls capture mice appropriately only if some part of their body moves. By contrast, birds, another common prey, are captured while they are entirely still; and as compared to mice they need to exhibit an elongate form tapering off in a tail that must not protrude rectangularly from the body (Räber, 1950). Thus a particular type of movement is an indispensable stimulus in one prey but not in another. (The nature of this correlation would seem adaptive. While mice are active in the night, birds roost and thus offer no movement to predators that are mainly nocturnal. It should be remembered in this context that, in the pigmy owl, birds and mice are associated with different internal states, but in this case the appropriate capture techniques differ slightly (Chap. 1. C. I.).)

d) If it could be shown that hypothetical afferent channels are differentiated by their relationships to different motivational systems, for example specific hunger (Chap. 1. C.), the multi-channel hypothesis would receive yet further support. As is well-known a number of reptiles and birds pick up grit to help grind the food in the gizzard or stomach. Snail shells might serve the same or some nutritional purpose and are taken up with typical capture movements by chameleons (O. v. Frisch, 1962; Brestowsky, pers. comm.) and anoles (*Anolis* sp.; Molle, 1957). Although the motivation underlying the ingestion of objects like these has not yet been studied, it is likely that it differs fundamentally from that underlying nutrition.

Chapter 4

Prey Selection

In selecting a particular individual of a prey species a predator may utilize cues that are much more subtle than those dealt with in the preceding chapter on prey recognition. Morphological cues involved may be so minute (e.g. Mason, 1965) as to shed doubt on the assumption that they offer the true cues for discrimination. But one has to bear in mind that these may be furnished by more obtrusive cues associated with them, for example behavioral ones.

A. Preying upon the Weak and the Sick

Field studies of large carnivores have provided most pertinent evidence for non-random prey selection. In a long-term study Pimlott et al., (1969) found that wolves in winter kill predominantly fawns and older age classes of white-tailed deer (Odocoileus virginianus) (Fig. 41). Deer killed more or less at random by automobiles or for research differ conspicuously from this age-class distribution. The percentage of fawns killed by wolves may actually be larger

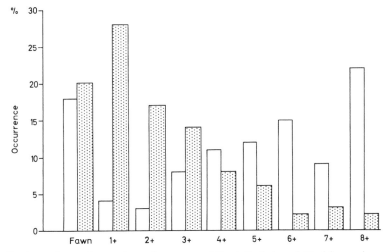

Fig. 41. Distribution of age classes (years) among white-tailed deer killed by wolves in winter (empty columns, n = 331) and by automobiles and for research (shaded columns, n = 275) in Algonquin Park, Ontario. (Redrawn from Pimlott et al., 1969)

than indicated because fawns are often more completely consumed as the skeleton is not yet fully ossified.

Although deer aged from one to three years composed 59% of the herd, only 15% of the deer killed by wolves were of these ages. A large fraction (68%) of the wolf-killed deer were four to eight years old, whereas animals of this age made up only 21% of the herd. Other detailed studies of wolf preying upon caribou (*Rangifer tarandus*; Kuyt, 1972), dall sheep (*Ovis dalli*; Murie, 1944) and moose (*Alces alces*) revealed likewise that young and old animals were killed in much higher proportions than occurred in the populations (review Mech, 1970, p. 249 seq.).

The reasons that disproportionately more fawns and old animals are killed may be diverse. The older a hoofed prey animal becomes, the more liable it is to exhibit various disabilities, as for example actinomycosis, hydatid tapeworm cysts, or depletion of bone marrow fat which is the last fat reserve to be metabolized. These conditions may well be interdependent and may add up in their effect upon the vulnerability of prey (see Mech, 1970, p. 258 seq.), and in buffalo-kills (*Syncerus caffer*) of African lions disease is inseparable from old age (Schaller, 1972, p. 232). Significantly, white-tailed deer killed by wolves were found to exhibit markedly more various abnormalities than a sample killed by hunters (Table 5). When considering this difference it must be borne in mind that determinations of physical debilities from the remains of wolf-kills represent only minimum estimates since abnormalities, for example, are often unilateral and thus escape detection (Mech, 1970). It is thus fair to surmise that fawns and calves do not suffer to any comparable degree. The relatively high vulnerability of these younger animals to wolf attacks would therefore seem to result from other reasons amongst which less stamina in fleeing is probably the major cause (see also Schaller, 1972, p. 317).

Table 5. Incidence of various abnormalities and pathological conditions in wolf-killed white-tailed deer, compared with that in hunter-killed deer. (Adapted from Mech and Frenzel, 1971)

	Wolf-kills		Hunter-kills		
Condition	Deer in sample	% Deer with condition	Deer in sample	% Deer with condition	P
Dental abnormalities	142	5.6	259	1.9	0.10 [a]
Jaw necrosis, lumps, or fractures	142	4.2	259	0.4	0.05 [a]
Pathology of lower limbs	75	6.7	126	0.8	0.05

[a] If both dental and jaw abnormalities are pooled, the difference between the incidence in the wolf-kill sample (9.8%) and that in the hunter-kill (2.3%) is significant at the 0.01 level.

A close parallel to wolf-predation upon the sick has been found in various birds (Fischer, 1970; Tamisier, 1970; Ullrich, 1971). Man-trained falcons (*Falco peregrinus, F. rusticolus, F. biarmicus*), when "thrown" at flocks of carrion crows, selected those that were debilitated in various ways significantly more often than a human hunter (40% vs. 21%). The difference would be even greater if one considers that the shot sample was biased towards the weak and the sick because these individuals would appear to be least suspicious of man (Eutermoser, 1961). Similarly, the incidence of injured or abnormally moving birds killed by four different raptor species amounts to 19.2 (14.3 – 33.3) %, a proportion exceeding the occurrence of debilitated birds in a population as a whole. It is difficult to assess how great the difference really is. The data available from an investigation of a sample of 10,000 starlings yielded an incidence of 5% "visible" abnormalities which may well be exceeded by less obvious ones (Rudebeck, 1950, 1951, and Refs.). Another case cautions that sickness per se may not directly enhance vulnerability but makes the prey move into places inviting attack. For instance, injured fish that still swim quite normally seek proximity of the surface and are thus more exposed to attacks from kingfishers than healthy specimens moving at greater depths (Kniprath, 1969).

Running predators such as canids during a chase experience the nature of the flight of the prey being chased. By contrast, stalking predators like cats do not have this opportunity. Accordingly, Schaller (1972, p. 395) and Kruuk (pers. comm. 1972) have suggested that coursers would select their victims by their physical abilities to flee and stalkers would tend to take a random sample from the prey population. The work of Schaller (1972) and Kruuk (1972 b) permits a test of this idea with respect to a single species preyed upon by both coursing and stalking carnivores. As can be seen from Table 6, African lions take Thomson's gazelle roughly in proportion to the population at large, if the latter sample is regarded as truly random. This finding would be in harmony with the above premise. But the data for three coursing predators, i.e. cheetah, wild dog, and spotted hyena, do not seem to support the idea; except for the two youngest age classes of gazelle, their toll corresponds by and large with the composition of the random sample. The reasons for the under-representation of age classes IV and V might be taken to support the original assumption, but, in addition, one would have to assume that flight capacities after that age remain constant, a highly unlikely ad hoc hypothesis.

The expectation is also not well borne out by the age composition of mule deer (*Odocoileus hemionus*) killed by puma and wolf, that is by a stalker and a courser, respectively. In a study area in Idaho, puma killed more adult bucks, calves, and fawns, and less adult does, than were present in the population at large (Hornocker, 1970). Bucks are believed to be less wary during the rut and exposed themselves more to predation by staying at the periphery of the herd or alone at higher elevations during the winter. In another study area in British Columbia puma were again highly selective with regard to age in that they took disproportionately more old (9 + years) animals of both sexes, disproportionately few yearlings (1½ years), and all other age classes at

115

random (Spalding and Lesowski, 1971). Apparently some physical disability as a result of aging rendered old deer of either sex more vulnerable than all younger animals. A similar conclusion can be drawn for glutton, which with respect to deer must also be considered a stalker (Myrberget *et al.*, 1969).

Wolf predation on the closely related white-tailed deer conspicuously parallels these results in that fawns and "old" animals (5 + years) suffered most and yearlings least, while other age classes fell slightly below expectation based upon a random sample (p. 114). A similar situation seems to exist in coyote predation on deer (Ogle, 1971) and pronghorn antelope (*Antilocapra americana*; Bruns, 1970). Fawns pay a disproportionately heavy toll though the old do not, perhaps because of being too large.

Thus again, despite vastly different hunting methods, puma and wolf, i.e. stalkers and coursers, took deer in proportions closely similar in terms of age. The same applies to selection in terms of sex when considering for puma the Idaho data solely. White-tailed deer bucks were taken by wolves in higher numbers than does though for a different reason. During severe snow conditions does have a lighter weight-load-on-track than bucks and therefore do not penetrate the snow as deeply, which permits them to run faster when pursued when wolves constitute a particular threat. Does have a weight-load-on-track that is only half that of bucks and thus increases their mobility in deep snow substantially. In addition, does are slightly less preyed upon by wolves the whole year round (Mech and Frenzel, 1971).

Apart from non-random prey selection by both puma and wolf in terms of prey age there is another parallel. As assessed by the bone-marrow method applied to a sample of 53 elk and 46 mule deer killed by puma, it does not cull selectively the undernourished (Hornocker, 1970). Likewise, among 69 wolf-killed deer there were only two individuals that had bone marrow depleted of fat stores (Mech and Frenzel, 1971). From these findings one must conclude that puma and wolf predation do *not* furnish evidence that supports the alleged stalker/courser dichotomy of prey selection. The selective culling of old animals by the two predators may, of course, have different reasons. As puma do not chase over long distances like wolves, the enhanced vulnerability of old animals must be assessed by much more subtle cues prior to a hunt.

Work with fish has remained equivocal. Pike preyed indiscriminately upon fantail and regular-tail goldfish, or upon differently colored individuals (Coble, 1973). The tanks in which the predation tests were performed may have provided so little cover that the results are not surprising, and in nature differential selection may well have taken place.

For an animal, being parasitized is often as harmful as being diseased or old. Endoparasites may render their host vulnerable in a number of ways. The parasite saps the stamina of its host, it makes it more conspicuous, or it affects the host's ability to respond adaptively to the predator by incapacitating sensory or central functions involved in escape (review Holmes and Bethel, 1972). It is in the interest of a parasite to maneuver its intermediate host into a position or condition which expose it to predation by the definite host. As demonstrated, cystacanth larvae of the acanthocephalan

Polymorphus paradoxus render the gammarid *Gammarus lacustris*, the intermediate host, vulnerable to ingestion by dabbling ducks (*Anas sp.*), the final host. The infested gammarids become positively phototactic, cling with their gnathopods to floating weed near the surface and when pushed off, "skim" along the water surface. By contrast, uninfested gammarids seek the dark, do not cling and, upon being disturbed, dive to the bottom. All three abnormal attributes lead to mallards (*Anas platyrhynchos*) ingesting the infested gammarids in predator tests, disproportionately often. Another species (*P. minutus*) renders the gammarid, in addition, more conspicuous by coloring it light blue (Hindsbo, 1972).

Numerous birds, while performing distraction displays, mimic a sick or lame individual ("broken-wing trick"). The widespread occurrence of this mimicry bears out that many vertebrate predators prey upon sick or otherwise disabled individuals if the opportunity arises.

B. Preying upon the Odd and the Conspicuous

General. Since long predation is thought to eliminate odd or otherwise conspicuous individuals from the population and thus to contribute substantially to uniformity. Group members are called odd if they differ by some feature from the majority. Individuals are called conspicuous if they do not match their background. Conceptually and in reality, oddity merges into conspicuousness if the group becomes an important part of the background (Fig. 42 a, see also p. 125). Doubtless the distinction between both attributes depends on how predators evaluate "background" in terms of camouflage and of spatial separation of individuals. If the odd individual is preyed upon more often, although it is better camouflaged than the majority of the group (Fig. 42 b) selection against oddity has almost certainly taken place. Conspicuousness has had little or no importance.

An examination of whether selection against oddity exists, meets in practice, two major difficulties. (1) When offering artificial prey of two colors Allen (1972) found that wild blackbirds took disproportionately more of the commoner form if the overall density was low (200 baits per 100 m²), and preferentially the rarer form if density was extremely high (300 in a 19 cm-diameter circle). In the latter case birds apparently treated the common type as a "background" against which the rare types stood out. Thus selection by blackbirds was directional (stabilizing) when baits were at high density and apostatic at lower density (but see Harvey *et al.,* 1974). Hence prey density needs to be carefully controlled in oddity experiments. (2) If predators tend to attack the deviant prey individual after having perceived the uniform group, comparison, if any, takes place in time. If the deviant individual is selected against, the response cannot be distinguished from a positive reaction to novelty. Predators that hesitate to attack novel prey and yet promptly take odd prey when separated from its group could provide the clue to an answer.

Because both issues have not yet been clarified, evidence for selection against "oddity" in the following can be merely suggestive. By comparison, evidence for selection against genuine conspicuousness appears well-reasoned (Trewawas, 1938) or irrefutable (Sumner, 1934, 1935; Isely, 1938; Popham, 1942; Dice, 1947; Brown, 1965; Tinbergen *et al.*, 1962; Zaret, 1972). A puzzling finding by Kaufman (1974) is that owls (*Tyto alba, Otus asio*) deprecate conspicuous white mice more than cryptic agouti mice, but curiously, in dense vegetation much more so than in sparse vegetation. The fact that loggerhead shrikes levied a heavier toll from agouti mice than from white mice if visibility was good, e.g. in sparse vegetation, and vice versa (Kaufman 1973 a), also remains unexplained.

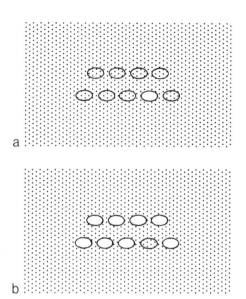

Fig. 42 a and b. Diagrammatic representation of two types of oddity in which odd prey contrasts to group members plus background (a), or to group members only (b)

Deviant Morphological Appearance. Goshawks select feral pigeons (*Columba livia* var. *domestica*) as prey, differing in color from their flock members; white pigeons were taken more often when most of the flock was black, and vice versa (Pielowski, 1959, 1961, further Refs. in Sulkava, 1964). Similarly, marine fish that had been marked individually with a fluorescent dye soon disappeared from their schools (Starck quoted by Hobson, 1968). Even minute details of the whole appearance, when changed, are selected against: wildebeest whose horns had been dyed white, fell prey to hyenas more quickly than normal animals (Kruuk, 1972 b, p. 154). In addition to these alterations the predator may have appraised behavioral abnormalities in the deviant individuals, though these were not obvious to the human observer.

Mixed-species flocks are assumed to offer a number of advantages (Chap. 2. C. IV. 2; Chap. 5. A. I.), but one has also to consider that a species forming a minority in such a flock may be regarded odd by predators and consequently be attacked first. For example, anchovetas (*Cetengraulis mysti-*

cetus) swimming in well-defined sub-schools amidst countless flatiron herring (*Harengula thrissina*) were preyed upon selectively by gafftopsail pompano (*Trachinotus rhodopus*). The anchovetas, being filter-feeders, swim with their gill covers widely extended so that these reflect the sunlight most conspicuously. From this Hobson (1968) infers that they acted as lures which drew the attention and directed the attack of the pompano. Likewise, kingfishers that have been observing a school of fish, immediately assume the strike stance, once a fish rotates around its long axis exposing its light underside like a flash (Kniprath, 1969). These observations touch on the problem of why schooling fish characteristically tend to school in groups of about the same size and similar external appearance (see Refs. in Roberts, 1972). As noted by Ehrlich and Ehrlich (1973) even heterotypic schools of reef fish that are composed of up to four species, tend to school in a way that reduces variation in body size. The most likely function of heterotypic schooling in this case, seems to be the protection from predators, as a number of other factors known to benefit schools can be ruled out. Heterotypic aggregations, as a rule, would then optimally have to consist of approximately equal proportions of different forms; any too odd minority might be jeopardized. Functionally, the same consideration would seem to apply to a polymorphism that can be discerned by a predator. Prey species should optimally have no morphs that are rare, unless a species lives solitary. But even for solitary species oddity or conspicuousness may be hazardous because of a predator's appetite for novelty (see Popham, 1942 below).

It may be added that advantages must be sought to exceed the advantage of forming schools of equal-sized individuals. In fact characids often do not school this way (Duncan, pers. comm.). Furthermore, when two size classes of "chub mackerel" were kept together, single smaller 3-in individuals mixed with schools of larger 7-in conspecifics but the reverse never happened (Parr, quoted by Welty, 1934). In the light of the facilitatory effect of group formation on social and other types of learning, the affiliation of small mackerel with larger ones may be interpreted as due to the benefits accruing to the younger fish, through tutoring by more experienced fish (Welty, 1934). On the other hand, a small fish amidst many larger fish is not as easily discovered by a predator as in the reverse case. Predator tests on this point are badly needed.

Movement. Oddity of movement ranging from slight abnormalities of locomotion through the virtual loss of control, as in panic flights, immediately catch the eye of predators and, as a rule, provoke instantaneous attack. In tests with rudd (*Scardinius erythropthalmus*) Popham (1942) found that when he removed part of one, or, in another trial, both hind legs, from corixids (*Sigara distincta*) these either swam in circles or swam more slowly than the normal ones that were present at the same time and in equal numbers. The experimentals of both amputated groups paid about the same toll, but a distinctly heavier toll than the normals (see also Cooke, 1971). As amputated and normal *Sigara* were presented in equal proportions and are equally non-schooling, "oddity" is likely to have been evaluated by a comparison of prey individuals in succession and would therefore not appear to differ appreciably from feeding on novel prey. A similar qualification applies to predation on "odd", equally non-social tadpoles. In wild populations of clawed frogs a certain proportion of tadpoles hatch crippled and thence swim abnormally and are eaten within a short time by adults inhabiting the same ponds. Savage (1963, quoted by Heusser, 1970, p. 390, in Grzimek 1970, vol. V) has speculated that by the maintenance of genetically defective cripples, adult frogs provide themselves with live food in a protein-deficient environment. However, according to Andres (1971, pers. comm.) laboratory hatchlings of *Xenopus* do not typically show such failures. The problem clearly needs further clarification.

By showing slight signs of "fright" induced by becoming separated from the group or by spotting an approaching predator, individuals render themselves vulnerable. This also applies to fish struggling (Banner, 1972), otherwise stressed (Herting and Witt, 1967) or disturbed (Potts, 1970; Markl, 1972; Zbinden, 1973) so that they induce the attack of piscivores, including squid (Neill and Cullen, 1974). Even slight disorientation of schools may render single members vulnerable to attack. As convincingly argued by Hobson (1968), disorientation may result from two schools meeting and rearranging their forward progression (see also Nursall, 1973).

When wildebeest were slightly disoriented by the strong lights of an approaching car or when recovering from narcosis, hyenas began to hunt them on the spot (Kruuk, 1972 b, p. 154); wildebeest attacked by larvae of a fly (*Gedoelstia* sp.) that occasionally penetrate into the brain, adopt a defective gait and other aberrations and are thus also more liable to predation

Table 6. Percentage of Thomson's gazelle of age classes IV – X killed by predators compared with ± random population sample. (Adapted from Schaller, 1972, Tables 48, 78, by courtesy of Univ. Chicago Press)

Class	Approx. age at beginning of class (months)	± Random sample [a] %	Lion-kills %	Cheetah-kills %	Wild dog-kills %	Spotted hyena-kills %
IV	9±	9	8.8	2.9	6.5	} 0
V	18±	17	11.6	2.9	6.5	
VI	24±	9	14.6	14.7	9.7	19.6
VII	›30±	31	33.6	35.3	45.2	} 42.9
VIII	»30±	13	8.0	11.8	6.5	
IX	»30±	9	8.8	13.3	9.7	} 37.5
X	»30±	12	14.6	19.1	16.1	
No. in sample		58	137	68	31	56

[a] Comprised of 27 animals collected by hunters at random, and of 31 animals found dead in a mass kill by hyenas in which selection for age can be assumed to be random. (Kruuk, 1972 b, p. 98)

(Schaller, 1972, p. 229). Fishes that flee or behave erratically draw the attack from a wide range of predators including *Octopus* with whom they coexisted peacefully a few moments ago. Murray eels (*Muraenidae*) and sharks appear to be particularly sensitive to fish in distress (Hobson, 1968). Similarly, wild dogs, cheetah, and hyenas immediately begin to chase Thomson's gazelle that have become panic-stricken, on seeing the approaching predator(s) or on being shot with a dart, instead of fleeing adaptively (Schaller, 1972, Kruuk, 1972 b, p. 100). Finally, birds such as shrikes (Ullrich, 1971) and jays (pers. obs.) which occasionally hunt songbirds, react in a flash to a small bird wriggling in a net; the stimulus is so powerful that they neglect a human standing close by. Birds of the crow family are attracted if a larger potential prey is going to be killed. Thus Kramer (1941) lured carrion crows by releasing a domestic chicken, recapturing it, releasing it, etc. Since crows are part-time scavengers their attraction to a place where a prey is being killed may mean that they anticipate to find the remains of it.

In quite a different context, genuinely odd individuals elicit attack, but this time from other group members. The attacks may inflict mortal wounds since the fights are not ritualized, as is usual between conspecifics (Wickler, 1967). The attacks often end with the expulsion of the deviant group member in humans ("*Ausstoßkämpfe*") or with its death in birds (Refs. in Eibl-Eibesfeldt, 1967, p. 331) and wild dog (Percival 1924, p. 48 quoted by Estes and Goddard, 1967). Since such hostility has only been reported among social animals, one may presume that it serves to eliminate a burden that would endanger cooperative behavior, which in the case of communal escape from a predator may become a hazard to more than the deviant individual. This view also commands support from the fact that the discriminating feature of the "outsider" must by no means threaten its health, as for instance a blotch of color on the plumage of a hen; but it may suffice to draw the attention of a predator, as will have become clear from the foregoing account. Thus intraspecific attacks, like predation, may attenuate individual variability, though for an entirely different reason. It should be mentioned that this process is not universal; it occurs in some highly social animals, e.g. some dolphins (Odontoceti) and wild dog (Schaller and Lowther, 1969), but has so far not been found in Mysticeti (Siebenaler and Caldwell, 1956; Pilleri and Knuckey, 1969). In some birds group members even care for debilitated fellow members.

Spatial Oddity. Individuals that stray from their group, i.e. are spatially odd, expose themselves to predation. The promptness with which predators take advantage of such individuals bears witness to the anti-predator function of group formation. This has been best documented by underwater observations of fish predation (Eibl-Eibesfeldt, 1962 b; Hobson, 1968). (Parting from the group may also have quite different disadvantages though (Chap. 2. C. II.).) Cuttlefish capture fish straying from the schools, though movement oddity seems to play a larger role (Neill, 1970). Similarly, hunting bats avoid patches where prey density is too high, perhaps because acoustical location of single flies becomes more difficult when others are within a short distance (Webster and Griffin, 1962). On the other hand, Webster (1967 a) adduced proof that capture success is neither impaired by targets close to the selected one and moving along similar trajectories, nor by clutter echoes; bats were able to capture one among 32 mealworms, or to separate a mealworm from an adjacent 3-mm sphere 2 cm away.

Predatory fish have evolved diverse tactics to strike at stray individuals. Some large marine piscivores (*Mycteroperca olfax, M. rosacea*) simply wait below a school of fish that is floating to and fro along the riff; as soon as a school member departs from the school they dart upward and snap it up (Eibl-Eibesfeldt, 1962 b; Hobson, 1968; see also Neill and Cullen, 1974: pike). Another technique is to "corner" the school communally, either by encircling it more and more closely (*Caranx adscensionis*), or by driving it into a shallow bay (*Carcharhinus* sp.) (Eibl-Eibesfeldt, 1962 b), a technique also used by fishing sea lions (*Zalophus wollebaeki*) (Brosset, pers. comm.); frightened individuals then dash away from the school and thus become spatially odd. A third tactic is to dart into the bunched school; in the ensuing commotion individuals become disoriented and segregate (Seghers, 1973), a tactic also employed by common loons (*Gavia immer*) when fishing (Allen, 1920 b), or by peregrine falcons (*Falco peregrinus*) hunting on bird flocks.

(Similarly, dense "air traffic" by large numbers of terns over a colony, facilitates fish robbing both inter- and intraspecifically; hosts cannot forestall attacks from parasitic terns when preoccupied with avoiding collision with other terns (Dunn, 1973 a)).

Why Prey in Groups Deter Predatory Attack. The first systematic investigation into the problem of how predators are deterred from attacking a group of prey has been conducted with schools of fish by Neill and Cullen (1974). These authors presented four piscivores, namely three professional ones (squid, pike, perch (*Perca fluviatilis*)) and a facultative one (cuttlefish), with singles, or schools of six and 20 prey fish. In nature, all four predators encounter part of their prey in schools, and in the laboratory all four were the more hesitant to attack, the larger the school (Fig. 43). Hunts on schools lasted longer, the larger they were. As successful hunting went on,

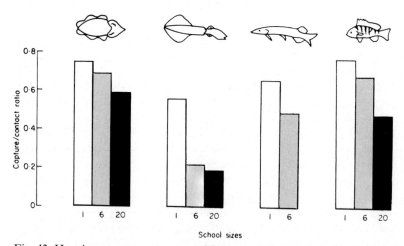

Fig. 43. Hunting success as measured by capture/contact ratios for four predators as a function of school size. From left to right: cuttlefish, squid, pike, perch. Contacts involve any discovery of a prey by a predator prepared to hunt. Further explanations see text. (From Neill and Cullen, 1974; by courtesy Zool. Soc. London)

schools became more wary and hence capture more difficult; and perhaps the predators were also satiated.

A sequential analysis of the predators' behavior while orienting towards and attacking prey revealed striking differences depending on school size. There were movements that were irrelevant in terms of hunting behavior, e.g. arms waving, protruding tentacles, V-posture in squid; and quivering, jerking, gaping, chomping, and S-bends in pike. Furthermore there were avoidance movements, e.g. ignore, hesitate, turn away, jet, ink, reverse in squid; and ignore, avoid, escape in pike. Both irrelevant and avoiding movements became more frequent as school size increased. By contrast, perch were more impeded by the frequent change of targets with the pursuer reverting to an earlier phase of the ideal undisturbed hunting sequence, as it switched to each new victim. Hence in three of four piscivores the school induced fear which, in conjunction with the activation of central prey excitation, apparently led to conflict behavior interfering with the final assault [7]. In perch, the fourth fish-hunter, schooling seemed to interfere more directly with aiming at a particular prey individual. While these behavioral changes which impair attack pertain to the predator, there may be others in the prey not considered by the authors; as is well-known, fish that habitually school, linger about "nervously" when separated from their companions. This behavioral change, which is obvious even to the human eye, may also invite attack from predators and may thus have enhanced the vulnerability of the individual.

Our own experiments (Milinski and Curio, 1975), using sticklebacks as predators of *Daphnia magna*, demonstrate a minimum group size above which a group enjoys protection as a function of numbers. What is more important is that any disturbance of the prey as a consequence of being single seems not to enhance vulnerability.

Experiments involving the selective choice between double prey objects by anurans suggest in which way a predator may become confused by a group of animals. When two equipotent stimuli are moved back and forth simultaneously on opposite sides of the midline within the binocular visual field, frogs hesitate to snap at either stimulus. Competition between both stimuli does not occur when they move in opposite monocular fields, or both within the lateral view of one eye. Competition is assumed to operate via a crossed-inhibition mechanism between rostral portions of opposite visual tecta (Ingle, 1973; see also Ewert and Börchers, 1970). Hence a predator could overcome competition between two prey objects by monocular fixation or by closing one eye. But such a method is not used, although monocular fixation does not impair the accuracy of the predatory strike, or the uptake of food in a number of vertebrates (Canella, 1963). Hence there must be other constraints that prohibit such a solution of the confusion problem. On the other hand, unilateral blinding leads to lowered capture success (56% vs

---

[7] There are indications that even small and harmless prey induce some fear in the approaching predator, not to speak of powerful prey that are attacked only under exceptional conditions (e.g. Leyhausen, 1973, pp. 27, 98).

90%) by cuttlefish presented with single prawns (Messenger, 1968). Since cuttlefish, too, are confused by shoals of prey (Neill and Cullen, 1974), it would be important to know if, after unilateral blinding, capture success in the shoal situation would improve.

Apparently, confusion by a school is not universal among predators: *Octopus vulgaris* when confronted with a mechanical "school simulator", i.e. with a "school" of baits, attacked it more readily than when offered a single bait, thus showing the reverse behavior of cuttles. Neill and Cullen (1974) interpret this difference by referring to the "blanket attack" of octopus, that catches a slow-moving crab by covering it with its web and arms. Cuttles are thought to aim more precisely because their swift prey is more elusive. However, it remains to be shown that an aiming mechanism with a greater margin of error becomes less upset by a group of targets than a more precise one. Another equally unproven possibility is that an octopus perceives a school as one large prey rather than as single members; its staple prey is solitary.

Selection against Oddity Versus Selection against Conspicuousness. In an attempt to separate the roles of oddity and conspicuousness, Mueller (1968, 1971, 1972) conducted predator tests with tame American kestrels and broad-winged hawks (*Buteo platypterus*). The birds were presented with live mice of two colors, each on a small pedestal with a distance between them of about 20 cm. The pedestals were arranged along the periphery of a circle with a radius of 2 m, in the center of which the hawk was perched. In one trial the hawk was allowed to prey upon one out of ten mice offered. One of the ten mice of each trial was odd, differing from the rest by its coloration only (grey, white). The pedestal background and floor were either the color of the odd mouse, or matched the color of the other nine. Trials were randomized with regard to the four possible combinations of the two color variants of mice and backgrounds. Tests with eight animals revealed that the odd mice were taken beyond and above their proportion in the total sample. But all eight subjects opted for odd mice of only one colour; four for white, and four for grey (Table 7). Only three of the eight birds took significantly more mice of the preferred color when these differed from the background, than when they matched it. Thus it seems that odd mice were preferred, but only when they were of a particular color, and that the conspicuousness of a mouse had relatively little influence upon the results. The preference for one color is explained in terms of searching image which is believed to be "the major determinant in prey selection in hawks" (Mueller, 1971, p. 345). But it is difficult to see why the birds should have taken the preferred color, according to their alleged searching image, less when it occurred more often than the non-preferred color; such was the case for four birds in situations *B* and *D* and for the other four in *A* and *C*, respectively. Just the opposite would have been expected had there been a searching image (Chap. 2. C.) so that the idea may be regarded as falsified. Mueller (pers. comm.) sometimes observed that the kestrels, when approaching a mouse, would land on a neighboring pedestal and thereby involuntarily come to perch upon the mouse there; from this he infers that the kestrel had not seen this mouse and must therefore have had a searching image for the one finally taken. However, I think that overlooking the neighboring mice may be an inherent feature of the target mechanism involved rather than evidence for a "searching image"; after leaving its perch the kestrel might focus upon the targeted mouse to the exclusion of scanning its surrounding. On the contrary, the bird cannot possibly have overlooked the other mice that it did not take, for then it could not have assessed that the one he took was odd. An assessment of oddity always involves a comparison between similar elements of the total situation. To this extent interpretations of the experiments in terms of searching image and of oddity are incompatible.

Table 7. Design of an experiment to examine the roles of oddity, conspicuousness, and individual preference in the attack of eight raptor individuals upon mice of two colors. In each experiment A – D with 50 replicates, odd mice were taken above and beyond their proportion (1/10) at a 0.05 level of significance: * 4 *Falco sparverius;* + 2 *F. sparverius,* 2 *Buteo platypterus.* For procedural details see the text. (Adapted from Mueller, 1971; Courtesy of Nature, MacMillan Journals Ltd.)

Experiment	Background	Mice presented in each experiment		Expected total catch if selection is at random	
		Grey	White	Grey	White
A	Grey	9	1*	45	5
B	Grey	1+	9	5	45
C	White	9	1*	45	5
D	White	1+	9	5	45

Mueller's (1971) other and more relevant conclusion that oddity is important for prey selection in his hawks has been challenged by Kaufman (1973 b), on the grounds that oddity may in fact have been mere conspicuousness (see already Meesters, 1941, Pielowski, 1961). The hawks may have perceived the non-odd mice as part of the background. Accordingly, the allegedly odd mice might have been simply conspicuous on a "prey background" but matching the painted substrate in two experiments (B, C) and on a "prey + painted substrate background" in two others (A, D). At present the question can neither be settled by pointing to the non-odd mice occupying only a minute fraction of the whole background, nor by pointing to their spatial separation from each other, as has been tried by Mueller (in Kaufman, 1973 b); such attempts will remain futile as long as it is unclear where the group ends and the background begins. Although the objection raised by Kaufman is undoubtedly correct, the possibility remains that the hawks used different mechanisms in different situations. They may have selected against oddity in B and C and against mere conspicuousness in A and D; the results obtained in experiments B and C would conform to expectations, had oddity been perceived (see p. 117).

Finally, it may be doubted whether the hawks used are the most suitable subjects for oddity experiments of prey selection. In nature, these raptors are seldom confronted with a simultaneous choice from a group of prey; as they travel along their searching paths they are more likely to encounter their prey in succession.

The objection against an oddity-centered interpretation raised above might also apply to choice experiments with three-spined sticklebacks and sunfish conducted by Meesters (1941). All three fish used in the experiments had been trained to snap up a bait, equally often, in front of two or three identical figures and an odd one. All subjects chose the "odd" figure in subsequent tests without reward. The strength of preference for "odd" worked regardless of which of two figures (triangle, rectangle) was the "odd" one. Still more remarkable is the fact that two of the fish chose "odd" after they had merely been trained to find food in front of a neutral figure (circle) which was not employed in the actual choice experiments. It is difficult to relate results such as these to the situation in nature but Meesters' experiments seem to come closer to a demonstration of selection against oddity than others; the figures protruded clearly against the background, so that the choice situation was less likely to be confused with camouflage.

As stated above (p. 121) predators tend to attack individuals straying from their group or split off by attacks. The question arises as to whether such individuals are genuinely odd, rather than novel, in which case appetence for variety of diet would also be a possible interpretation. Whether or not stray individuals would appear attractive because of being novel to a predator, would depend on how often individuals would become separated from the group. If this occurred regularly, novelty would be expected to give way to familiarity and hence to a change of the intensity of selection against stragglers. This problem clearly awaits further investigation.

Counteractions of the Prey. The predator's approach, or its attempts to separate individuals from the group, are commonly counteracted by bunching which builds up resistance either automatically, or through confusion. Clusters of an aphid bunch upon being threatened by a hymenopteran parasite. In this way the parasite is prevented from penetrating into the middle of the cluster (Fig. 44), so that only the marginal individuals remain vulnerable (Klingauf and Sengonca, 1970). The same holds for the feeding behavior of an anemone-eating nudibranch (Waters, 1973). Fish (Allen, 1920 b, Eibl-Eibesfeldt, 1962 b; Potts, 1970), tadpoles (Black, 1970), birds (Tinbergen, 1952; Horstmann, 1953; Brooks qoted by McKinney, 1965, p. 287; Ainley, 1974), ungulates (Bruns, 1970; Mech, 1970; Schaller, 1972), and banded mongoose (*Mungos mungo*) (Kruuk 1972 b, p. 274) also bunch upon sensing a predator, or upon being attacked, though species differ markedly: while thrushes and tits are poor at bunching together, rooks, jackdaws, and long-tailed tits are good (Meinertzhagen, 1959, p. 23). Upon pursuit, mullet (*Mullus* sp.) schools may split in two and swim away in opposite directions and so may add to the confusion of the predator (Neill and Cullen, 1974; see also Pilleri and Knuckey, 1969; Nursall, 1973). The tight formation flights of black-headed gulls responding thus to a flying predator, yield the same effect by performing sudden and synchronized turns (Kruuk, 1964, p. 33 seq.), i.e. a "protean group display". The mode of escape and its direction may vary with the species of the predator.

The separation of an individual from its group may be mitigated by its greater wariness. For example woodpigeons when alone, look up more frequently than flocked birds (Murton *et al.,* 1971). Similarly, groups of elk (*Cervus canadensis*), moose, and mule deer have a lesser flight tendency than solitary (or peripheral) individuals (Altmann, 1958). In a quantified case of predation the benefit derived from such great wariness seems to accrue solely to groups of minimal size. Schaller (1972, p. 257) believes that smaller herds of ungulates (Thomson's gazelle, wildebeest, zebra) are more wary than larger

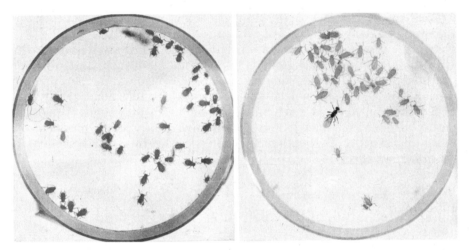

Fig. 44. An undisturbed colony of the aphid *Aulacorthum circumflexum* on synthetic food (*left*); upon approach of a hymenopteran parasite (*Diaeretiella rapae,* Aphidiidae) the aphids congregate. (From Klingauf and Sengonca, 1970)

ones. To what extent solitary individuals benefit from their greater wariness is unknown; they fall prey to lions at a higher rate than herds of any size (Schaller, 1972, p. 446).

As pointed out by Williams (1966) and Hamilton (1971) on theoretical grounds, schooling can confer an advantage for a prey animal even if it does impair the predator's capture performance. Because any prey individual that moves into, or finds itself in the center of a school (group), will be less liable to predation than individuals at the periphery. This holds, though the group as a whole would not be off better. Thus this sort of protection may well be understood in terms of individual selection rather than group selection (Williams, 1966). However, the consideration does not apply universally, for some predators do not hesitate in the least to plunge in the midst of a dense swarm, as for example a *Cichlasoma* cichlid chasing young fish (*Tilapia zillii*) (pers. obs.), or peregrines feeding on bats flying in streams from their cave (Meinertzhagen, 1959, p. 153). Milinski's experiments (unpublished) in our laboratory, using sticklebacks as predators, show that such behavior depends on how hungry these are.

In Conclusion. There are many casual observations and some experimental evidence, that predators, when given a choice select prey individuals that differ from the rest of their group; whether this evidence will hold for a comparison of prey in time, remains to be seen. Differences involve morphologic appearance, movement, and spatial oddity. In some social animals natural selection acting against deviation from the norm is exerted by conspecifics, probably to reduce the vulnerability of the social community as a whole. There is good evidence from fish and cephalopod piscivores that a group of prey deters predators from attack because it induces fear and/or interferes with the functional sequence of hunting in some other way. Reasons are discussed, as to why aiming, that always seems to involve binocular fixation, is upset by a group of prey (confusion effect). Oddity is a special case of conspicuousness in which the odd member stands out from its group as a matching background. Experimental evidence as to whether selection against oddity actually takes place is equivocal and can be more parsimoniously, though not entirely, explained by selection against conspicuousness in terms of camouflage. Bunching of prey animals and the greater vigilance of small groups are effective anti-predator mechanisms. Being a member of a group already confers an advantage, independent of any confusion effect.

C. The Mechanics of Prey Selection

Any selection against the old or the weak can be accounted for by two different explanations that are not mutually exclusive. Either the weeding out of these individuals is an inevitable consequence of the mechanics of the hunt, in that they are less wary or slower than the healthy segment of the population and therefore automatically fall prey to the attacker (Mech, 1970, p. 262). Or the attacker makes its selection before it attacks or starts the chase by detecting signs of weakness in its would-be quarry. Unfortunately, there are no direct data, that permit a discrimination between these two possibilities; the proportions of healthy and weak individuals in successful and unsuccessful hunts remain as yet to be assessed. However, there are two indirect pieces of evidence of a true choice on the side of the predator.

a) Hyenas approach herds of wildebeest in their ordinary walking posture or gait. On being close, they often dash into the herd, and after a short run, stop and watch the fleeing herd, and then select the individual they are going to chase. From this revealing observation Kruuk (1972 b, pp. 149, 164) infers that hyena "test" by their short initial run, the flight abilities of all herd members and would then select their quarry after signs so subtle that they do not become obvious to the human observer. Wild dogs select their quarry similarly (Estes and Goddard, 1967; H. and J. van Lawick-Goodall, 1970, p. 61; see also Murie, 1944; Pielowski, 1961).

In a different manner, a pack of wolves that hunt moose "test" a would-be victim by dashing in at it. A healthy moose stands at bay on the spot or after a short run. The majority, however, including the weak, bolt off and continue to flee with the pack closely following. A moose standing its ground and belligerently charging at intervals at the nearest wolf will generally be harrassed for less than a minute (Fig. 45). The wolves have probably learnt that they cannot overpower such an animal unless it is wounded or sick; in this case the pack may keep watch on the encircled animal for hours and eventually kill it. Another method of "testing" consists of making large herds of caribou run and then single out a weak animal lagging behind (Murie, 1944; Mech, 1970, p. 205 seq.). Similarly, cheetah often begin a hunt after Thomson's gazelle at submaximal speed and accelerate only after the gazelle have begun to run and the cat has presumably made its selection (Schaller, 1972, p. 317; see also Kruuk, 1972 b, p. 189). It must be realized that the

Fig. 45. Wolves "testing" a moose that stood its ground until after 5 min the wolves gave up and left. (From Mech, 1970, p. 219)

whole argument involves the danger of the reasoning going round in circles. Ideally, the observer should have evidence concerning the physical condition of the prey apart from the choice he sees the predator exert.

b) The length of a chase is determined by the quarry's condition. African lion chase after Thomson's gazelle fawns over much longer distances than after adults, apparently because during the chase they sense the prospects of the hunt (Schaller, 1972, p. 247). In another case the caution noted above (a) can be met with: A wolf hunted down a deer with an arthritic hind foot over four miles whereas chases after healthy deer do not exceed some 100 yards. The wolf had probably noticed during the chase a defective gait in its quarry (Mech and Frenzel, 1971).

c) Lanner falcons when hunting often pursue one of the most distant individuals in a flock of carrion crows and ignore those closer and within deceptively easy reach (Eutermoser, 1961; see also Meinertzhagen, 1959, p. 55, *Larus marinus*, p. 98, *Melierax metabates*).

d) The existence of selection against oddity constitutes unequivocal evidence of a choice, if only in time, prior to a hunt.

A predator that concentrates upon a vulnerable quarry is obviously at a greater advantage than one that squanders energy and time by hunting an animal that can outlast it. Age-selective predation by puma raises the question of whether a stalking predator also, with no opportunity to test its prey, can make its selection prior to the final lunge. From spoors in snow, an unusually high (see Schaller, 1972, p. 455) success rate (82% of 45) of the final attack upon deer and elk could be assessed (Hornocker, 1970). This seems to preclude the idea that individuals killed were actually selected at random. Accordingly, the perceptual powers of assessing the vulnerability of prey by puma appear extraordinarily good.

D. Evolutionary Implications

From the foregoing, it has become clear that predators eradicate the weak and otherwise conspicuous (odd) individuals of a prey species. Selection against genuinely odd individuals is further enhanced by social hostility directed against them. Despite these threats, a number of selective advantages are accrued by the rare phenotypes, that ensure their maintenance in the species and thus promote variability. The necessary frequency-dependent selection may take a number of forms. (1) It may be apostatic, i.e. conferring an advantage on the rarer of two phenotypes because it alleviates the formation of a searching image in the predator (Chap. 2. C. VI.). (2) A rare Batesian Mimic will beat an advantage over a common one, if the associated models are only moderately distasteful; as the rare mimic increases in frequency it loses its advantage, for the predator then becomes increasingly liable to discover the difference between model and mimic and will no longer be deceived by their similarity (Clarke and O'Donald, 1964; O'Donald

130

and Pilecki, 1970; Pilecki and O'Donald, 1971. (3) Non-random mating was shown to confer considerable selective advantages to the rarer of two genotypes in males of *Drosophila*, a phenomenon now known as the Petit-Ehrman-effect (see Parsons, 1973, p. 78 seq.). Again, as the rare male genotype increases in relative frequency its mating advantage disappears and the formerly common genotype male gains in advantage up to an equilibrium state for both. The behavioral mechanism underlying this rare mate advantage, is only just beginning to be explored (Ehrman, 1970) while in other cases of non-random mating it is as yet entirely obscure (e.g. Sheppard, 1952).

In higher vertebrates, frequency-dependent selection that would maintain rare phenotypes is expected to be of much less significance, because Batesian mimicry here is rare. Moreover, the incidence of the rare mate advantage is certainly complicated by the preponderance of sexual imprinting.

Slobodkin (1968, 1974) has suggested that predators should "manage" their prey so as to insure its continued availability. This "management" should alter the pattern of non-predatory mortality as little as possible. The optimally "prudent" predator should take those prey that have the lowest reproductive value and thus contribute least to future generations. Since predators cull the young and the old, or those that are parasitized and have consequently become sterile (review Holmes and Bethel, 1972), predators indeed appear "prudent". It is widely accepted that they weed out the expendable excess (e.g. Errington, 1946, Jenkins *et al.*, 1963, 1964). Yet at the same time the young and the old are those that are easiest to get. Therefore Slobodkin's (1974) view that predators have contributed to reproductive value distribution as a function of age is difficult to test. It might be put to a test with a prey species in which the lowest reproductive values do not coincide with the highest vulnerability in terms of age.

Chapter 5

Hunting for Prey

A. Modes of Hunting

Prey species differ in their defense tactics, if only subtly. Accordingly, many predators preying upon more than one prey employ a number of hunting methods while others are committed to a single hunting pattern for life. This, together with the fact that the diets of most predators differ from each other, points to the existence of almost as many hunting repertoires as predator species. Yet some generalities have come to light.

I. Hunting by Speculation

A number of predators "hunt by speculation" in that they behave like a blind man feeling his way or searching for objects with his stick. *Octopus cyanea* regularly makes speculative pounces upon coral rocks, clumps of algae, or a small area of the bottom on its foraging trip. Every one to two meters it quickly closes its web over the target, followed by the lateral arms, bases, tips, and central parts of the leading arms in this order (Fig. 46). The

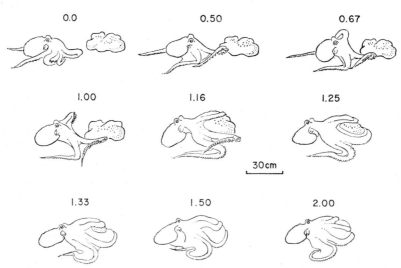

Fig. 46. *Octopus cyanea* making a speculative pounce on a coral rock from a crawling position. *Numbers:* time in sec from beginning of the sequence. Drawn from 16-mm film shot at 24 frames/sec. (From Yarnall, 1969)

132

target area next to the leading part of the body is covered last and is surrounded by the arms and then covered by the web comprised of the entire interbrachial web and the free membranous parts of the arms. Pausing for a few seconds to feel under the web, the octopus continues its trip. Captured crabs are immediately paralyzed and carried to the home to be eaten (Yarnall, 1969).

Lions that are intimately familiar with their hunting range, at times run up a slope to peer down from its ridge, clearly suspecting a vulnerable prey on the other side. Or, a lion trots toward prey and only speeds up once an animal has proven inattentive; the speculative and unobtrusive trot brings the lion closer to a prey than any speedy, and hence warning run. On top of that, the greater momentum of trotting as compared to walking facilitates any attack that develops from the trot (Schaller, 1972, p. 245 seq.).

Predators relying predominantly on visual hunting must rely on tactual and/or taste stimuli when searching for prey in a medium which impairs or precludes vision. Thus storks (*Ciconia ciconia, C. nigra; Mycteria americana*) find submersed prey in water so turbid or choked with vegetation that vision is impossible (Löhrl, 1957; Kahl and Peacock, 1963). A wood stork (*M. americana*) typically inserts its partially open bill into the water, moves the bill slowly from side to side, and walks forward. When a fish makes contact with the bill, the mandibles are quickly closed, with a reflex time of 25 msec (20 – 32 msec). Experimental restriction of the frontal vision in one bird, depriving it of at least the central 75° of visual field, scarcely impeded fish capture. Such non-visual prey capture would enable an animal to feed in the night as well (Kahl and Peacock, 1963). Similarly, marsh mongoose (*Herpestes paludinosus*) grope for animal prey with their forepaws, like raccoons, under the water surface (Grzimek, 1972, p. 173, Vol. XII), a behavior which has also been reported for spotted hyenas trying to relocate a submersed carcass in a pond or river (H. and J. van Lawick-Goodall, 1970, p. 167; see also Leyhausen, 1973, p. 51). Various larids and a wader (*Totanus melanoleucus*) are known to "plough" or "skim" the water with the lower mandible immersed cutting the surface, seizing any prey contacted; in ploughing the bird runs through shallow water in the manner described, in skimming it flies just over the surface, the obligatory feeding method of the skimmer (*Rhynchops nigra*) (Tolonen, 1970; see also Dunstan, 1974), though a cleaning function may also be served in some species (Buckley and Buckley, 1972). The various scratching movements performed by gallinaceous birds and a number of songbirds (Davis, 1957; Curio, 1959 b) belong essentially to the category of hunting by speculation.

Many aquatic birds use foot-stirring in various ways to pick up subsequently prey flushed or whirling about. Herons while hunting stir up debris with one foot vibrating in or above the surface of the mud or raking it, and spear fish fleeing (Meyerriecks, 1959). The stirring foot vibrates at a rate of 12 – 13 per sec (Helbig, 1968). A number of waders employ various one-footed vibrating (*Vanellus, Charadrius*), jumping (*Tringa, Calidris*), or trampling movements (*Scopolax rusticola*), similar to those of gulls (discussion Tinbergen, 1958), in order to make prey animals emerge from ground not covered by water; in addition, these waders can already locate prey that is still underground by their faint sounds (Lange, 1968)[8]. By tapping the ground with one foot a lapwing (*Vanellus vanellus*) made earthworms appear (Fallet, 1962). To the group that use foot-stirring must be added flamingos (Phoenicopteridae), storks (Ciconiidae), and the hammerhead stork (*Scopus umbretta*) (Rand, 1956).

---

[8] While Lange (1968) claims that a ring ouzel (*Turdus torquatus*), too, can locate subterranean prey by sound, Heppner (1965) has shown that hearing, does not necessarily play a role in the capture of earthworms by the American robin (*Turdus migratorius*).

A slightly different variant is used by owls (*Strix aluco, Tyto alba; Micrathene whitneyi*) which beat upon branches of trees to flush birds from their roost, or insects (Refs. in Scherzinger, 1970; Ligon, 1968). Yet Smeenk (1972) doubts that, for instance, tawny owls can catch birds in mid-air. However, even if they cannot, startling a bird at its roost may give the owl a chance before the prey takes wing. Further, it still needs to be examined if the varied wing and tail flicks of small passerines commonly exhibited between flights, serve to flush insects as suggested by Root (1967). Similarly, Brosset (1969) believes flycatchers (*Tschitrea batesi, Diaphorophya castanea*) "whip" the foliage with a conspicuously long tail and employ a noisy wing-flapping to the same end, when wandering in mixed species flocks. However, the near-to universal species-specifity of such movements (Andrew, 1956; Curio, 1959 a) would suggest that their main function is somewhat different.

Another common variant of the flushing of prey is the following. Cuttlefish (Tinbergen, 1966, p. 14) and three groups of puffer fishes (Krapp in Grzimek, 1970, p. 250 seq., Vol. V: Balistidae, Tetraodontidae; Longley, 1927 quoted by Hobson, 1968: Ostraciodontidae) have the habit of blowing a jet of water, either with the funnel or with the mouth, respectively, towards the sandy bottom and thereby expose or flush prey animals, e.g. small crustaceans hidden in the substrate. [Balistids also use the blowing movement when preying upon sea urchins by turning them over with a powerful jet so that their oral side becomes vulnerable (Fricke, 1971)].

Some marine fishes and birds take prey flushed by other individuals, even of other species. For instance, goatfish (Mullidae) forage in the sand digging about with their barbels, often burrowing their snouts up to their eyes. According to Fricke (1970; see also Hobson, 1968) species of at least eight families of fish accompany goatfish intently regarding the spot disturbed and periodically dart forward to snatch up organisms uncovered (Fig. 47). The feeding goatfish raises a cloud of sand, and as experimentally shown, such a cloud will attract fish to within a distance from where they watch with caution. If a mullet dummy is moved through the cloud the allured commensals lose all caution and will even contact the dummy as they usually contact the live mullet. The latter alone elicits the following behavior prior to any actual feeding by the mullet. Apparently, the follower species have learned to associate with prospects of a meal, both the sand cloud and the mullet (Fricke, 1970).

There are numerous examples of birds that make use of ants, other birds, or of mammals as "beaters" moving through the herbage and thereby flushing insects otherwise remaining hidden. The relations between the foraging raids of army ants (e.g. *Eciton burchelli, Labidus praedator*) in the American tropics and birds, for instance the ant-tanagers (*Habia* sp.), are well known descriptively. The birds follow the columns of ants through the undergrowth low over the ground, peering down onto the pillaging mass and snatching up flushed insects (Willis, 1960; 1972).

Cattle egret ride on large browsing mammals or accompany them on foot, or, as in Jamaica following mowing machines instead (pers. obs.). Similarly, carmine bee-eaters (*Merops nubicus*) ride upon Abdim's storks (*Sphenorhynchus abdimii*) or kori bustards (*Ardeotis kori*) and swoop in pursuit of insects disturbed by the mounts. The use of monkeys as "beaters" has been described for such different birds as hornbills (Bucerotidae) in Africa, drongos (Dicruridae) in Asia, fairy bluebirds (Irenidae) in the Philippines, and trogons (Trogonidae) in Panama (Hartley, 1964). That it is not only the disturbed insects which attract birds to the beaters is shown by an observation by Brosset (1969). Hornbills in Gabon which regularly follow foraging parties of various monkeys (*Cercocebus albigena, Cercopithecus nictitans, C. mona, C. cephus*) have been attracted by caged individuals in the absence of any flushed prey insects.

Fig. 47. A wrasse (*Halichoeres nicholsi*) closely follows a feeding goatfish (*Mulloidichthys dentatus*) to feed on organisms flushed from cover as the goatfish disturbs the sand. (From Hobson, 1968)

Still commoner are mixed species flocks that move through e.g. neotropical forests in search of insects in the absence of any larger species functioning as beater. It has been suggested that within a mixed feeding flock each bird acts unwittingly as a beater for the other members. Some species called "catalyzers" initiate by their particular behavior (continual pirouettes, wings hanging limp, tail spread as a fan) flocking behavior of many species called the "followers". "Catalyzer" behavior occurs in several Asiatic cuckoos, at least one American tyrant flycatcher and several of the American tanagers (Brosset, 1969). The cohesion of flocks may be enhanced, for example in mixed species flocks of Central American forests, both by similar color and behavior of the party members (Moynihan, 1962, 1968). Besides offering feeding advantages flocking behavior certainly serves a number of other functions as well, among which protection from enemies seems to be the most prominent (e.g. Moynihan, 1968, Willis, 1972; see also Chap. 4. B.).

II. Stalking and Ambushing

1. Stalking

If the sustained speed of a predator does not permit successful attack, an unrecognized or mistaken identity means everything for approaching toward the unsuspecting victim until within range for the sudden assault. This active approach is termed stalking, whereas ambushing leaves the approach to the prey. Both methods may be exercized by the same predator. The general coloration of stalking predators is concealing and thus supports their stealthy approach. While stalking, tell-tale body movements are avoided and advancement is slow. Open-water stalkers, e.g. cuttlefish (Neill and Cullen, 1974) or longnose gar (*Lepisosteus osseus*) (Foster, 1973), approach their victim by inconspicuous undulations of transparent membranes or membranous fins.

Another method in stalking, is the extremely skillful use of cover, used by, for example, pike (Benzie, 1965), or the larger cats when working up to their game (Cott, 1957, p. 143). Lion and cheetah advance in just the precarious moments when the prospective victim has lowered the head to graze, or stands facing away (see also Benzie, 1965: pike), "except in the case of large herds when it might be futile to wait for all animals to be inattentive" (Schaller, 1972, pp. 247, 316). When an animal suddenly becomes alert the lion halts, sometimes standing motionless with paw raised in midstride. When the prey moves behind some bushes or out of sight into a ravine, the lion may run closer.

An important attribute of the stalk is an adaptive reduction of apparent body size and of noise. Thus the stalking predator crouches or moves head-on so as to minimize visible movement. Functionally related is the search flight low over the ground of accipitrine hawks and the attack flight of great grey shrikes during which they make use of even slight depressions and can hence hardly be seen as an alarming silhouette (Cade, 1967; Glutz et al., 1971). While trotting towards a herd of wildebeest, wild dogs lower their heads and bunch, an action which usually lowers the flight distance of the prey (Schaller, 1972, p. 329). A similar function may be attributed to their traveling in single file while going on a hunt (H. and J. van Lawick-Goodall, 1970, but see Estes and Goddard, 1967).

Adaptive silence is important to outwit prey that can hear. Predators such as the stalking carnivores avoid any noise detrimental to their hunting activities. Owls are equipped with various feather structures that render their flight almost inaudible. A corroboration of the significance of these structures is the existence of owls that hunt after prey insensitive to the sound of flapping wings and that do not possess these structural modifications; the fishing owl and the elf owl, for example, have no need of a silencing device, as their prey, being under water, or being arthropods, respectively, are (almost) unable to hear the noise of their wings (Graham in Cott, 1957, p. 145; Ligon, 1968).

It is an old question whether stalkers mask their scent by approaching prey up-wind. Lions have been credited with this propensity by Selous (in

Cott, 1957, p. 146), yet Schaller (1972, p. 240) has made it clear that wind directions during 300 hunting episodes varied at random. If wind direction plays a role in the hunting of other predators, it would have to be accompanied by the ability to make detours while approaching the prey. A variety of predators are able to make detours while stalking which involve a temporary retreat from a closer to a farther position, followed by a subsequent new approach, thus demanding considerable control over hunting

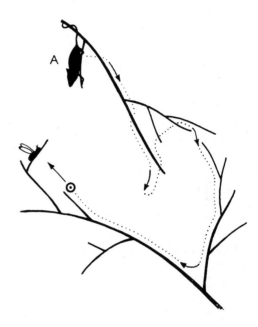

Fig. 48. Solution of a detour problem by a pigmy chameleon stalking a winged insect. A = point of discovery; $\odot$ = point of tongue shoot. (Modified from O. v. Frisch, 1962)

behavior (Fig. 48). Detours are solved by young and adult pigmy chameleons (*Microsaurus pumilus*) in the same manner; the animal advances until within optimal strike distance (O. v. Frisch, 1962). In addition to this chameleon the ability to perform detours has been well documented for pigmy owl (Scherzinger, 1970, p. 34), for puma (Hornocker, 1970), and for lions encircling their prey during communal hunting (Chap. 5. C. III. 1.). Four out of five wolves pursuing a deer ran around a lake shore as if to intercept it after it had sought refuge by swimming across the lake; the fifth wolf swam after it, while the others probably killed it after landing (Mech, 1970, p. 226). Salticid spiders have been credited with "long detours" (Crome, 1957). In view of its uniqueness in invertebrates a closer study of salticid hunting behavior appears highly desirable.

In general, during a detour, visual contact with the quarry is interrupted. This situation is reminiscent of searching for prey that has disappeared from the predator's visual field (Chap. 2. B. I. 3.). The important difference lies in the fact that during an active detour visual separation is taken into the bargain as an inevitable outcome of hunting tactics, whereas in the search situation it is not.

Stalkers are usually capable of at least a short run, to bridge space, without which even the greatest stealth would be doomed to futility. Or a swift dash is made after prey as it bolts off if the actual assault has failed. Which mode of approach is chosen, depends upon the speed or the manner of movement of the prey, and upon some subtler properties of the situation as a whole. With a prey that moves slowly, praying mantids (Holling, 1966), salticid spiders (Drees, 1952), and cats (Cott, 1957, p. 143; Leyhausen, 1973) stalk as unobtrusively as possible and thus minimize the risk of untimely discovery before the final strike or rush. Furthermore, dragonfly larvae, like lions, stalk prey that has ceased to move with even more stealth than slowly moving prey (Hoppenheit, 1964 b), presumably because still prey is more apt to detect movement in its surroundings. By contrast, prey with rapid or jerky movements that might herald rapid disappearance, is approached quickly until within striking distance, by all the above-mentioned predators. Obviously, the predator is thereby taking a chance, if only a small one, instead of losing an opportunity [9]. Accordingly, both quick movement of the prey as well as dull prospects of a hunt make a stalker run and thus probably maximize its chances of capture.

2. Ambushing

An ambush predator tries to conceal or advertise its presence while lying in wait for a prey to come close. Concealment is achieved by camouflage and suppression of visible movement (see preceding chapter) and/or by a well-chosen hiding-place. If advertisement is used it inadvertently takes the form of aggressive mimicry (Chap. 5. A. III.).

On spotting prey a number of ambush predators exhibit "discovery movements" that might disclose the predator's presence and thus unfavorably interfere with hunting success. Their magnitude may be adaptively related to the nature of the prey. Upon prey detection cuttlefish show "running on the spot" with their velum, yet to a lesser degree when approaching vigilant fish, than the less vigilant crabs or prawn (Neill and Cullen, 1974). Cats and certain geckos twitch their tail tips after detection of a prey. These movements have been alluded to, as acting as a lure, distracting the attention of the prey from the deadly forepart of the stalker, and thus facilitating its hunt. An illuminating case concerns an adder (*Bitis peringueyi?*) of the Namib Desert. It lies in ambush for small birds so that only the eyes and the tip of its tail protrude above the sand surface. The tail tip is moved when a poten-

---

[9] It is no chance coincidence that *escape* tactics may be similarly structured. Ptarmigan (*Lagopus* sp.) crouch upon seeing a gyrfalcon at close range and they take wing well ahead of the enemy when it is still far away. If ptarmigan find themselves on bare ground and the falcon is too close as to make rapid escape flight feasible they *slowly* move to a patch of cover in order to crouch there (White and Weeden, 1966). A similar behavior has been observed in dall sheep (*Ovis dalli*) rams approached by wolves when these had not yet begun to run (Murie, 1944). This firm control of escape behavior parallels the slow stalk of the approaching predator in that quick progression is effectively suppressed.

tial prey animal is spotted. Curiosity makes the latter approach the tail tip and thus come within striking distance (Sauer, pers. comm.). Since the tail tip could easily be hidden beneath the sand, as is almost all of the body, the tail movement may be interpreted as a prey-catching device rather than as an accidental outflow of "central prey excitation". However, it should be stressed that such a lure hypothesis in these and other cases is absolutely unproven.

Various ambush predators position themselves so that they clearly see a prey as a silhouette against a light background. Thus sharks (Eibl-Eibesfeldt, pers. comm.), groupers (*Mycteroperca, Epinephelus*) (Eibl-Eibesfeldt, 1962 b; Hobson, 1968; Neill and Cullen, 1974), and nightjars (Schlegel, 1967) dart upward to snap off prey that contrasts with the light sky. Ambushing may be aided by various auxiliary devices such as webs, sticky threads, traps or pits. Actively closing snares seem to be employed only by certain fungi for capturing nematodes (Pramer, 1964) with fast death being achieved by the excretion of ammonia (Balan and Gerber, 1972).

Surprisingly, one of the most obvious questions concerning selection by an ambush predator of an optimal ambush site, has not received the attention it deserves [10]. A lurking predator such as a spider in its web, an ant lion (*Myrmeleon*) in its pit, or a harpactocorine reduviid (Heteroptera) waiting for insect visitors to flower heads may or may not get enough food. If prey remains too scarce an ambush predator will rather move to a more favorable site to wait anew than starve to death. An illuminating case in point is provided by Łomnicki and Slobodkin (1966, see also Ritte, 1969) who reported that *Hydra littoralis*, a typical ambush predator, drifts off with the aid of a gas bubble generated beneath the pedal disc if prey becomes too scarce, in order to take a new chance elsewhere in the pond. The floating response is mediated by a "conditioning" compound emitted from neighboring fellow *Hydra*. Responsiveness is highest at moderate population densities (8/10 cc); below and above them a much smaller proportion of the *Hydra* will react. Water that has been "conditioned" by starved *Hydra* will elicit less floating than water from well-fed *Hydra*. Łomnicki and Slobodkin (1966) interpreted the density-dependence of the floating response as structured in such a way as to make use of the past history of a colony: an area containing a high density of individuals would be an area where prey is or has been abundant, so that a too-rapid departure would be a strategic error in the face of a sudden deterioration. If the crowding level is not very high at the time of food supply deterioration, the prognosis for the location is not as good, and the advantage of departure is correspondingly greater, but at low densities competition is at its lowest and the benefit of departure smaller.

Ambush predators often prove markedly versatile in selecting an optimal ambush site, though reports are little more than anecdotal. A captive anglerfish (Antennariidae) gave up lying in wait on the ground when prey stayed away; it straddled to a new position beneath the surface from where it

---

[10] A paper by Turnbull (1964) on the cobweb-weaving theridiid spider *Achaearanea tepidariorum* is often but erroneously believed to adduce the desired information.

snapped off coursing "minnows" without ever operating its baiting tassel (Illicium) (Reese, pers. comm.). African crocodiles (*Crocodilus niloticus*) position themselves at strategically optimal places, such as riverine passages where fish have to pass, that are hence "shovelled" at times (Guggisberg, 1972).

Raptors of several species (mostly *Buteo* sp.) after waking up in the morning, first indulge in a "warming-up" flight of usually less than an hour. Thereafter they stay for 3 – 4 h in each of two hunting areas within their home range. At the utmost they remain perched for 1 – 2 h in any one place while lurking for prey, and change perch up to 52 times per day; a single perch may be revisited 2 – 3 times a day (Craighead and Craighead, 1956, p. 47).

Clearly, sessile animals not able to leave their home site must in a way be able to predict critically, before settling for life, the chances of thriving. That such a prediction must be extremely difficult is indicated by the fact that the majority of permanently sessile animals are filter-feeders, with a virtually constant food supply (Cloudsley-Thompson, 1972). A multitude of stimuli serve as orienting cues for larvae of barnacles and of serpulid polychaete worms during settling on the species-specific substrate (review Reese, 1964).

III. Prey Attack under Disguise

General. Stalk and ambush predators try to minimize their visibility until within reach of the final assault or spurt. Quite a different method is to make the prey, or host for that matter, mistake the true identity of the predator and, in doing this, to rely upon distinctly visible signals. The disguise involved has become well-known as aggressive mimicry or Peckham's mimicry. At first glance aggressive mimicry appears totally different from concealment or camouflage during stalking. However, as convincingly argued by Wickler (1968, p. 238 seq.), both concealment and types of mimicry (except strictly Müllerian) involve deceiving the signal receiver, in the present case the prey or host. While this is undoubtedly true, concealment and mimicry impart quite different demands upon the signal emitter. Hence the distinction of both tactics still appears useful if the research focuses upon the predator rather than the prey. Reports on methods of disguise are largely anecdotal and hence must be taken as no more than starting points for a more penetrating analysis.

Methods of Disguise. Both chasing and ambush predators have developed means to disguise their whereabouts or true identity from would-be prey. The chasing predator may either use a harmless animal as "magic hood" or it possesses itself structural and behavioral devices rendering it similar to a harmless or beneficial species. Using an especially subtle sort of disguise the predator does not deceive the prey as to its appearance but "pretends" not to be "interested" in the prey. It lingers close to it, without any sign of stealth. In this way piranhas (3. A. II.) and Norway rats (Steiniger, 1950) manage to launch an attack from close quarters upon fish and birds, respectively. In still another way pipe-fish (Aulostomidae) (Eibl-Eibesfeldt, 1955 a, Hobson, 1968) and Pacific cornetfish (Fistulariidae) (Hobson, 1968) make use of parrotfish (Scaridae) (Fig. 49), groupers (Serranidae), or mullets (Mullidae) as a "mount". A parrotfish when approached by a pipe-fish tries to evade it but the latter follows stubbornly and skilfully and, as time goes by, will be tolerated (Eibl-Eibesfeldt, 1955 a). From their mount they slide away sideways and snatch, with a lightning strike, suitable prey fish that happen to come too close, taking advantage of the indifference of small fish that do not fear the usually harmless mount; they swim away just in front of its mouth and thus become vulnerable to the pipe-fish. It would be interesting to

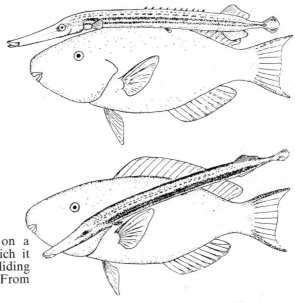

Fig. 49. Pipe-fish "riding" on a parrotfish (*above*) from which it attacks small fish after sliding slowly off its "mount". (From Eibl-Eibesfeldt, 1955 a)

know whether prey fish are not startled, because they overlook the riding pipe-fish, or whether this is due to the unobtrusive movement of the mount.

Aggressive mimics may gain a double advantage from their deception. The blenniid fish (*Aspidontus taeniatus*) imitates the cleaning wrasse (*Labroides dimidiatus*) by its coloration and by the dance the latter employs when approaching a cleaning client. By its disguise *Aspidontus* succeeds in getting close enough to strike and inflict wounds in the fish that invite cleaning from the cleaner. *Runula rhinorhynchus*, a slightly less perfect mimic of the cleaner, has been found to be recognized by victims that have experienced its attacks. Previous victims learn to discriminate it from its beneficial model *Labroides*. Hence the discriminative power of wounded fish places a selective premium upon the mimetic resemblance of *Runula* and other aggressive mimics like *Aspidontus*. Since the cleaner is safe from attack by larger fish its mimic enjoys the double advantage of getting at its unsuspecting victim and of being protected against potential predators

141

(Wickler, 1968, p. 162 seq.). The same applies to a salticid spider (Eisner *et al.*, 1969).

There are many inspiring guesses as to the true nature of aggressive mimicry in other cases. For example, a cooper's hawk (*Accipiter cooperi*) has been seen to fly like a jay (*Cyanocitta* sp.) when approaching unsuspecting sparrows (Peeters, pers. comm.), and merlins (*Falco columbarius*) fly frequently in the fashion of small passerines when being about to attack them (Rudebeck, 1951, and Refs.). Further, one of the two color morphs of the European cuckoo has been credited by Löhrl (1950 b) with a deceptive resemblance to the kestrel (*Falco tinnunculus*) in color and both are thought to resemble sparrowhawks (*Accipiter nisus*) in flight pattern. Verification of these and similar observations seems almost beyond the scope of quantification.

A unique hunting technique has been described by N. G. Smith (1969) for two Panamanian forest falcons (*Micrastur mirandollei, M. semitorquatus*). These raptors attracted, on altogether five occasions, small birds and toucans (*Rhamphastidae*) by clearly species-specific calls. Lured native birds remain hidden in the foliage and "mob" the vocalizing falcon, the calls of which are difficult to localize, though much more so in *M. mirandollei*. The falcon suddenly dashes off to catch one of the lured mobbers. North American winter visitors are similarly attracted by the ventriloquial calls of the falcons. However, they neither remain hidden nor do they utter mobbing calls [11] as the native species do and therefore presumably offer easier targets for the falcon's attacks. Both types of vocalization by the falcon serve a puzzling double function: because of their high pitch they are difficult to locate and yet they make prey birds congregate around the sound source. Still more enigmatic is the reason why birds are lured by calls which are associated with immediate danger.

Recently, Wilson, (1972, p. 366) described deception by chemical means in ants. Slave-making ants (*Formica subintegra*), when raiding in colonies from which slaves are robbed, spray substances (decyl, dodecyl, and tetradecyl acetates) from their enlarged Dufour's gland at the defending colonies. These substances are alarm pheromones of interspecific significance that have been termed "propaganda pheromones" because they help to alarm and disperse the defending workers that thereby are confused and search for the wrong enemy. The raiding workers themselves are not affected, and thereby have their way paved free for robbery (see also Hölldobler, 1973).

Man is the only aggressive mimic amongst mammals. African natives affix a stuffed head and neck of the ground hornbill (*Bucorvus abyssinicus*) to

---

[11] These "high-intensity alarm cries" are easy to locate and seem to have been confused by Smith (1969) with the high-pitched aerial (i.e. flying) predator calls that are difficult to locate (Marler, 1957). From the description given, it is clear that the former vocalizations are mobbing calls given when mobbing a perched owl or hawk. (Whether a flying hawk really locates prey birds by their vocalizations is still a matter of speculation. An observation by Stoddard (1961, in Brown and Amadon, 1968, p. 31) on a cooper's hawk seems to support it; an indirect support is the very existence of high-pitched aerial predator calls).

their forehead while they are stalking towards their prey (Fig. 50). The ground hornbills are common birds to which the game animals, exploited by man, are entirely accustomed and from which they need not fear any danger. Similarly, North-American Indians use wolf-hides when approaching bison (*Bison bison*), and Australian aborigines, kangaroo skins when hunting kangaroo. The latter hunters also mimic normal browsing, locomotion, and comfort behavior that all add to the disguise (Wickler, 1968, p. 120). When using duck decoys that attract wild ducks, or when using the well-known plethora of artificial baits for angling, man is betraying his game animals in still other ways.

Fig. 50. Hausa bowmen in Northern Nigeria hunt prey under disguise of stuffed ground hornbill headdress, a stalking horse technique known, from cave paintings, to be thousands of years old. (Redrawn from a photograph in Fisher and Peterson, 1964, by courtesy of The Mansell Collection). *Inset figure:* a live ground hornbill

Various other methods of disguise have not been documented beyond the level of anecdotes. Angling devices by which prey can be baited have been evolved a number of times. The tassel (Illicium) of the angler-fishes (Lophii-formes) or worm-like structures that are also moved like worms in a catfish (*Chaca chaca*) and the alligator snapping turtle (*Macroclemys temminckii*) are too well-known to deserve closer description here. While these ambush pred-ators allure their prey with part of their body, three individual green herons (*Butorides virescens*), related to each other, had learned to utilize an object other than their own body for that purpose. In a unique way they mandibulated and thus dissolved commercial fish food pellets in shallow water of the Seaquarium in Miami. They thereby attracted small fish which they readily snapped up and ate (Sisson, 1974). The performance is noteworthy because the food pellets cannot have served as the unconditioned stimulus, during the acquisition of the habit. Incidentally, the general procedure of those herons is reminiscent of tongue-flicking (Buckley and Buckley, 1968) or bill vibration movements with which snowy egrets (*Leucophoyx thula*) attract mosquito fish on which they then feed. Bill vibration does not attract four other species of prey fish and is applied by some individuals more than by others, though the hunting success by this capture technique is comparable with three others commonly used (Kushlan, 1973).

An undocumented case of prey lure with food, has been reported for a tame chimpanzee (Denis quoted by Kortlandt, 1972, pers. comm.). It had learned, without tutoring, to lure sheep by means of food, to seize them, tear them apart, and eat them.

It is still unknown whether feigning death by animals has been developed for any other purpose than protection from enemies. Red fox have been alleged to attract crows by feigning death; when they came close enough, mistaking the fox for carrion, it would catch them with a quick dash (Müller-Using in Grzimek, 1972, Vol. XII). A more likely though equally unsubstantiated case concerns the piscivorous cichlid fish *Haplochromis livingstonii* of Lake Malawi. It is supposed to sham death by lying on its side on the sandy bottom. The color pattern, a pearly lustre on a ground color of dirty white, together with the unusual posture, produce the impression of a fish which is not only dead but in the early stages of decay. Fish attracted by the conspicuous "carcass" are presumably suddenly seized when within strike distance (Fryer and Iles, 1972, p. 207).

IV. Pursuit of the Prey

A chase may be preceded by lying in wait and/or by careful stalking. The prey's chances of escape will depend on when in the functional sequence leading to a full chase the predator is detected. The point of discovery by the prey will influence in turn, amongst other factors, whether the predator will take up a chase, continue it or give up. The predator may also give up when the quarry is out-distancing it or does not fall back soon enough. The distance over which a chase continues may be influenced by a number of factors apart from the speed of the quarry: lions chase after young antelope, other things being equal, for longer than after full-grown ones (p. 130). A chase ending in capture is by no means always due to the maximum velocity of pursuit, as will become clear in the next section.

1. Changes of Velocity of Attack (Pursuit)

a) *Changes of velocity in relation to control of attack* have been found in cinematographic analyses of the hunting behavior of cephalopods (Maldonado, 1964) and fish (Nyberg, 1971). *Octopus vulgaris* individuals were familiarized with capturing a crab tethered to a bar at a fixed distance from their home. Upon switching on a light the animal began to approach the crab and accelerated its forward progression in a characteristic way (Fig. 51): at first there was a phase (second time delay), during which the animal accelerated, decelerated, and also approached with almost constant speed (zero acceleration). During the subsequent strike phase (final pattern of acceleration) acceleration and deceleration attained peaks, but never fell to zero; moreover, this phase was in general shorter than the initial phase. In experiments

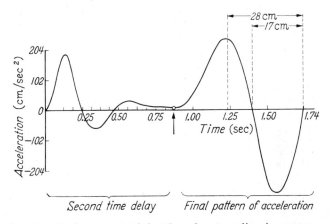

Fig. 51. The curve of acceleration during an attack by *O. vulgaris* ending in capture of a tethered crab. During the first phase of attack (.88 sec) the animal covers only 18% of the total distance; during the strike phase, the onset of which is marked by a dot on the curve and an ↑, it covers the rest, with the whole movement time being 1.74 sec. In the strike phase the positive maximum of acceleration is reached 28 cm in front of the crab; the deceleration begins 17 cm in front of it. (From Maldonado, 1964)

designed to discover how much of the whole attack movement was under external visual control illumination was interrupted at various points of the cycle. When the octopus had viewed the crab slightly in excess of the first phase before the light was switched off it attacked with a strike pattern of acceleration, that did not differ from an attack which was fully illuminated throughout. Since during the strike phase, the attack does not depend on virtually continuous estimates of the distance from and the location of the goal, an internal control mechanism must produce a program of forces before the strike phase begins. Thus the distinction between the first and the final attack phase is not only justified in terms of the pattern of acceleration, but also on the basis of the nature of control of attack.

In a similar study using the largemouth bass (*Micropterus salmoides*) Nyberg (1971) found that there is also a point during the attack, at which a change occurs from sensory (visual) feedback control to an internal control, whose output is not modified by subsequent sensory information. After or

145

just before the bass begins to open its mouth, to suck in and grasp the prey, it no longer modifies the direction of its attack path. Prey capture will be most successful when the fully-opened mouth contacts the prey. The time required to open the mouth is about linearly related to bass body size and not to velocity. Therefore the faster a bass travels the greater the distance covered during the time needed to open its mouth. Hence the distance from the prey at which the bass begins to open its mouth must be determined by an estimate of its velocity. This needs to occur at the latest when the internal control mechanism takes over. A comparable chaining of external and internal control mechanisms has also been found in the attack of mantids (Mittelstaedt, 1957) and cuttlefish (Messenger, 1968). Likewise, kingfishers close their eyes as they hit the water surface in their dive after a fish (Eastman, 1969), and owls prior to impact upon the prey animal (Norberg, 1970). The adaptive significance of this sort of cut-off is probably protection of the eyes.

b) Velocity of attack may depend on the *nature and location of the prey*. In the largemouth bass relative attack velocity in mid-water is 3.1 body lengths per sec and is independent of predator size. In a given individual it varies more during attacks in mid-water and at the surface than during attacks of prey located on the bottom. In which way these variations of the approach towards prey change the probability of capture remains unknown, except that the final stage of the approach depends on the type of prey. In the capture of "minnows" the bass' velocity increased during the .02 sec before the jaws began to open; by contrast, if the prey was a worm (in mid-water), i.e. a passive prey, the attack decelerated in five out of ten feeding episodes, remained constant in four others, and increased in speed in one other case. Prey on the bottom was invariably approached with a slowed-down attack and often the bass stopped completely. Such prey were crayfish (*Orconectes limosus?*) or worms. Hence, the bass seems to respond adaptively to the potential mobility of the prey it is going to seize. An increasing velocity would increase the probability of capture, if the prey was timing its escape response on the basis of a velocity estimated earlier. With less mobile prey such an increase would be unnecessary.

Another cinematographic analysis of bass that were preying on zebra danio (*Brachydanio rerio*) revealed that the reactive distance of the prey increased with increasing approach velocity (Dill, 1973). Upon the prey fleeing, the bass speeded up its approach ("pursuit velocity") until within striking distance. The sooner a prey detected the approaching bass the longer was the distance from which the latter elected to strike and hence tried to compensate for the prey's greater "head start". Longer strike distances, however, seemed to result in fewer captures. Thus prey-reactive distance may have an indirect effect on strike success though this is not fully borne out by the data. Both approach velocity and pursuit velocity increase significantly with increasing experience, but not with hunger, as measured by the number of previous captures. It is difficult to understand why greater experience should indirectly (by means of a longer strike distance) lead to reduced capture success as just suggested. Various hypotheses are being tested to account for

146

this discrepancy (Dill, pers. comm.). Maldonado (1964) found that in *O. vulgaris* strike distance increased with experience. Strike success decreased as strike distance increased in octopuses of little experience, but less so in more experienced ones. It should be noted, however, that the crab was stationary whereas the bass' prey was free to flee. Clearly the intricate relationships between the parameters discussed need further experimental elucidation.

c) Cheetah may begin a hunt at moderate speed, presumably to "try out" the most promising victim, speed up maximally when having selected one and slow down again. This deceleration allows the cat "to *follow each twist* in the unpredictable route of its quarry. With danger imminent, the gazelle may zigzag but seldom does so more than 3 or 4 times" (Schaller, 1972, p. 317). At other times the cheetah bounds toward an unsuspecting herd at full speed whereupon the herd flees precipitately and without dodging.

d) Linear progression during pursuit may also become slowed down when prey is "*cornered*" and its panic movements in front of the pursuer prohibit precise aiming. "Cornering" of prey has been reported for perch, trout (Benzie, 1965), shark (Eibl-Eibesfeldt, 1955 a, Eibl-Eibesfeldt and Hass, 1959), sea lion (Brosset, pers. comm.) and for zebra fish (*Pterois volitans*) where it is aided by widely spread and poisonous pectoral fins (pers. obs.). It may be added that another obvious constraint on full speed in the "corner" and related situations is that a predator may get hurt when hitting the substrate in lieu of the prey. Larger piscivores when hunting inshore occasionally become stranded, in the fury of their attacks (Hobson, 1968). Open-water hunters are not endangered in this way but can be carried by the momentum of their attack beyond the point of prey seizure, taking along the prey without any harm (Messenger, 1968; Nyberg, 1971). By contrast, effective stopping must be exercised right after impact by various raptors and owls; the bird curtails postimpact horizontal movement which could be clearly hazardous in foliage, by thrusting its wings forward, dropping down its tarsometatarsus, and leaning forward (Goslow, 1971).

e) Linear progression may also become slowed down in that *triangulation* when aiming at prey is facilitated by a nonlinear track, though this possibility still needs confirmation. For example, royal terns (*Sterna maxima*) dive onto the surface in a spiral-shaped track and both cattle egrets and bustards (Otididae) perform a damped horizontal oscillation with their heads when picking up prey in front of them (Buckley and Buckley, 1974).

f) In some cases flight speed of prey stays below the highest possible pursuit speed. A zebra family, for example, with one or more foals that *cannot* gallop as fast as the remainder of the group flees at moderate speed. Spotted hyenas then follow easily trying to separate a foal from the herd, with the stallion falling back to defend the family (Kruuk, 1972 b, p. 184).

Protrusion of body parts employed in the actual seizure of the intended prey accelerates velocity in the last and most critical moment before impact. Raptors (Fig. 52) and owls (Payne, 1962; Norberg, 1970) then thrust their feet forward, so that their speed exceeds that of the pelvis, e.g. by 15% (equivalent to 2250 cm/sec) in the goshawk. The pelvis in turn is thrust forward 30% more rapidly than the head which is moved back while the wings and tail develop maximal brakage (Goslow, 1971). Seizure of the prey by the bass is aided by rapid protrusion of the premaxilla when the jaws become fully opened. Since protrusion velocity is constant for a given bass, the additional velocity gained by protrusion would seem to be more important to an attacking bass traveling at low speeds than to one moving at high speeds. Furthermore, for young bass this additional velocity is of greater importance than for older ones because swimming velocity of attack increases with size at a faster rate than velocity of protrusion: For 25-cm bass, protrusion makes up 50% of the attack speed, whereas for 10-cm bass the respective velocity gain amounts to as much as 87% of average attack velocity. Apparently this size relationship is the functional reason why large and powerful swimmers

Fig. 52. Typical striking posture of goshawk behind a pigeon. Tracings made from 16 mm film with frames spaced at 30 msec intervals. (From Goslow, 1971)

that capture fish at very high speeds, like barracuda (*Sphyraena* sp.), have lost protrusibility (Nyberg, 1971).

A high speed of attack confers a double advantage to the predator. It not only helps to surprise the prey, it may also make school members overlook an attack. Thus schools of minnows (*Phoxinus phoxinus*) or mugilids did not notice when several of their members had been captured by a cephalopod or fish predator. In this way the school is less quickly alarmed by the attack than might be expected (Neill and Cullen, 1974).

2. Interception of the Flight Path

Any potential prey moving out of alignment with the predator, affords it the chance to draw closer by intercepting the prey's path. In intercepting, the pursuer makes a prediction about a future location of the prey. It is unknown on what information this prediction is based. Likely sources of information are the prey's direction and presumably its velocity in relation to the predator's velocity. Interception has been noted to occur presumably in fish (Dill, 1973), and certainly in birds hunting highly mobile prey (e.g.

Angell, 1970; Hatch, 1970, 1973; pers. obs.), in bats (Roeder and Treat, 1961), wolf (Murie, 1944; Mech, 1970, pp. 211, 226; Mech and Frenzel, 1971), lion, and wild dog (Schaller and Lowther, 1969; Schaller, 1972, pp. 250, 340).

In wild dogs, during the chase after a quarry, the pack spreads out behind the leader who initiated the chase. Hence a quarry that veers off from the straight flight path or, after a hard chase of 1 or 2 miles, begins dodging, often falls prey to a pack member other than the leader. The initial spreading of pack members thus facilitates later, involuntary shortcuts (Estes and Goddard, 1967). As a matter of fact, intentional shortcuts by pack members do occur, when the quarry begins to circle (H. and J. van Lawick-Goodall, 1970, p. 60). With one or two dogs following a gazelle and most of the others cutting across the arc the victim suddenly finds itself surrounded. Brief relay hunts may ensue, with one dog taking the place of another that had come to the fore (Schaller, 1972, p. 340). As affirmed by H. and J. van Lawick-Goodall (1970), the lead dog may increase speed as soon as one of the others swerves off, to cut corners; similarly in wolf hunts, excepting those due to mechanical constraints of the hunting grounds, a single file with wolves following each other is much more common (Mech, 1970, p. 211 seq.).

The fanning out of the pursuers may have the additional effect of confusing the prey. Hamilton (1973, p. 122) has suggested that the piebold, coloration of communal hunters, such as wild dog and killer whale (*Orcinus orca*), which hunt in schools (Martinez and Klinghammer, 1970), serves to signal the prey the location of its many pursuers and so add to their confusion. This may be true and the question as to why other communal hunters, e.g. spotted hyenas, do not have an advertising coloration may be nothing but another demonstration of the ubiquitous opportunism of evolution.

Hamerstrom (1970, p. 105 seq.) witnessed many prey captures of a tame golden eagle (*Aquila chrysaëtos*), some of whose tactics revealed considerable skill. When chasing over plain country after its very first hare, the eagle stooped down in mock attacks without seriously attempting to seize the hare. During these mock attacks the hare commonly dodged. Just before reaching cover it was caught by the raptor, i.e. in a situation when dodging was of no avail without considerable risk of being hurt from the vegetation. Another tactic was a puzzling kind of interception: The eagle overtook the bounding hare, when it had backwind, and when the hare no longer felt pressed, the eagle sharply twisted around and caught it unsuspecting (see also Gordon, 1955, p. 96). The opposite has been observed in some species of cats that habitually attack a prey from the rear. If the prey animal happens to face them they perform a somersault over its back, twist around right after landing and leap onto the prey's back (Leyhausen, 1973, p. 106). It goes without saying that the sharp turn in front or behind the prey, respectively, serves different purposes: While the cats by their somersault get into a position from which they most easily leap onto the back of their victim and avoid its teeth at the same time, the function of the unusual maneuver in the eagle is difficult to understand; its superior speed would have allowed seizure from behind anyway.

Unlike interception described so far, young herring (Rosenthal, 1969 b), pike, and *Coregonus* larvae (Braum, 1963) do not anticipate the direction of the moving prey copepods, as judged from their path of advancement. When aiming at one, they rather bend their body in such a way as to hit the prey, despite its movement across the visual field (Fig. 53). While binocularly fixating the prey a herring larva maneuvers itself by undulating movements of the pectoral and the dorsal fins into striking position so that the head is

Fig. 53. Posture of herring larva (18 mm body length) just before striking at prey copepod. *Arrow* denotes swimming direction of prey. Note position of dorsal and pectoral fins. Flashlight picture. (Modified from Rosenthal, 1969 b)

pointing roughly perpendicular to the overall path of the prey. At the same time the body is generally bent in a direction opposite to the swimming direction of the prey (Fig. 54). As a result, the body unbends in attack so that the head lunges in a direction that is displaced sideways and opposite to the preparatory flexing of the body. In this way the forward progression of the prey becomes intercepted. Strangely, pike and *Coregonus* larvae, as documented by film analysis (Braum, 1963), bend the body in the direction of the later head displacement which also coincides with the swimming direction of a prey copepod. The preprogrammed strike is again an adaptive anticipation of prey movement. Why interception is achieved both by body bends of opposite signs remains so far unexplained in hydrodynamic terms.

It is noteworthy that herring larvae bend forwards and backwards when preparing to strike at a molluscan veliger larva. These larvae habitually perform rotating movements and any systematic displacement of the strike direction would be futile. The same switching of the preparatory bend occurs, presumably for the same functional reason, when older larvae fixate a copepod that swims in their own direction or towards them.

150

Although not documented for pigmy chameleon, this ambush predator also seems to anticipate the prey's path when "shooting" (O. v. Frisch, 1962). Yet it is unknown whether it extrapolates movement or anticipates it from an evaluation of the prey's front-tail-orientation. This latter clue is, however, utilized by herring larvae that anticipate the potential future location of a prey by its front-tail-orientation even though it does not move (Rosenthal, 1969 b). That is, they bend in such a way as if the prey would move during

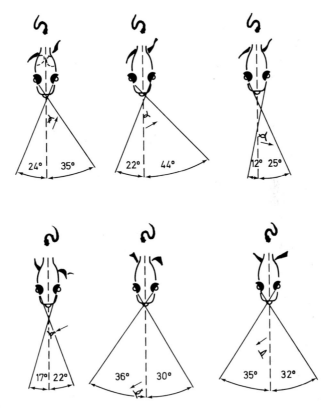

Fig. 54. Fixation of prey copepod just before strike by herring larva (cf. Fig. 53). Note body bend, position of pectoral fins and eye fixation movements as a function of direction of prey movement. Drawn after flashlight pictures. (Modified from Rosenthal, 1969 b)

the strike and thus demonstrate that they discriminate between the front and the tail region of the prey. Similarly, cuttlefish maneuver sideways or behind a prawn or crab; in this way the attack with the specialized pair of tentacles intercepts the prey's expected or actual backward dart and the chelae are also avoided. Conversely, fish are approached from the front, the direction in which they would be expected to flee (Messenger, 1968; Neill and Cullen, 1974). Furthermore, largemouth bass approach a crayfish by swimming slowly past its front and stop for attack when the tip of the jaws reaches the back of the prey's tail. On the other hand, the attack on fish in mid-water is rather precipitate, at least under aquarium conditions. The bass also bases judgement of which part of the prey's body to hit in attack on mechanical constraints. It seems to aim at some point in the middle of the prey, if it is fragile and elongate like worms; and onto the nearest point if it is a fish (Nyberg, 1971). To what extent predatory tactics vary with the type of prey will be discussed more fully below (Chap. 5. B. I.).

Another way of interception, forestalling the prey's escape, is practised by polar bears (*Ursus maritimus*) when hunting seals on pack ice. After detecting specimens resting on the ice, the bear swims towards the edge at a point where the escape route of the seal towards the open water would be shortest. In the final stretch of approach the bear swims submersed to prevent untimely discovery. It then leaps onto the ice and thus cuts off the quarry (Pedersen in Grzimek, 1972, p. 135, Vol. XII). A related method is employed by lions anticipating the direction of traveling prey by placing themselves in ambush on the track ahead; this tactic is also an integral element of their communal hunting (p. 200).

In Conclusion. Predators base their interception of the actual or potential flight path of their prey on properties of the prey's movement and on its front – tail orientation, that indicates the potential route of escape; and on its precise location in relation to an escape route. Interception may take the form of shortcuts during a chase; of placing an ambush site alongside a game track ahead of the traveling game or ahead of prey that are made to flee; of aiming the strike ahead of the moving prey or ahead of the potential flight direction when judgement can be based only upon morphologic cues and the prey is stationary.

3. Counteradaptations of the Prey

Velocity of Flight. Prey species capable of rapid flight are often swifter than their predators. African lions can reach maximum speeds between 48 and 59 km/h when running after a quarry; with few exceptions, however, their hoofed prey animals can run faster (Schaller, 1972, p. 233). Similarly, a sprinting man stays just below the maximum sprint and sustained speeds of many potential prey animals (Hamilton, 1973, p. 126). Comparable estimates for invertebrates pertain solely to an ambush predator. Roeder (1968, p. 143) assessed the strike of the praying mantis (*Hierodula*) to last approx. 60 msec, after which time the forelegs have grasped a fly. It is noteworthy that the flight (escape?) response of flies, i.e. the interval between take-off from the substrate and the moment they fly, takes some 55 msec, as does the startle response of cockroaches. The latency of the whole escape response, including the response time of the visual system would, obviously, last somewhat longer and thus give the mantis a good chance. It is not farfetched to assume that in the past predators have been influencial in shaping these speed relationships.

Protean Displays. Startled or chased prey often escapes attack by "protean displays" involving zigzag movements, twists, loopings, etc. These maneuvers engender unpredictability of the escape route and hence upset the target mechanism of the pursuer, which is often unable to follow by correspondingly rapid changes of its path of pursuit. Moreover, protean displays obviate an improvement of capture tactics through learning, since learning is based on minimal predictability (review Humphries and Driver, 1970). Relatively few predators have overcome the protean defense. It has been documented that bats (*Myotis lucifugus*) use their flight membranes to enlarge the area of the open mouth effectively to catch insects. The bat localizes a flying insect acoustically from a distance of 25 – 80 cm and attempts to intercept the flight path. Most commonly bats (Vespertilionidae) make use of their tail or interfemoral membrane (Uropatagium) to ensure firm seizure of the prey by the mouth; the membrane is pouched by forward flexion of the hind legs and tail (Fig. 55 a). But the wing membranes are also used, e.g. when the

Fig. 55. (a) a typical mealworm catch by the bat *Myotis lucifugus*. *B2 – B5* and *M1 – M3:* positions of bat and mealworm during consecutive flashes that are separated by .083 sec. At *B4* the tail membrane is being snapped forward completely over the face, and the mealworm is grasped at one end between the bat's teeth, as can be seen in *B5*. (b) a greater horseshoe bat (*Rhinolophus ferrum-equinum*) catching a moth (*Catocala* spec.) with the aid of the left wing. Flashes .15 sec apart. This species does not use its tail membrane to aid in insect capture. (Modified from Webster and Griffin, 1962)

bat rolls to its side, to force down an insect that is flying too high in the moment of interception, or to scoop an insect with the bent tip of one wing from where it is transferred to the mouth (Fig. 55 b) [12]. In bending the wing tip the terminal joints of both third and fourth fingers are flexed to reach the mouth. In vespertilionid bats it may be "first retrieved with the wing tip and then either be carried or flicked into the pouch of the tail membrane" (Webster and Griffin, 1962, p. 334). From the pouched membrane the prey will then be picked out by the mouth. There are hardly any catches without preparatory maneuvers of the approaching bat. These maneuvers are to assure capture even if the trajectory of the prey is almost linear, as in the case of mealworms "shot" into the air. The quickness with which a final wing-tip action is achieved is in excess of 1 cm/msec and the precision of the final placement approximates a volume roughly within 1 ccm of the required location. Since this capture action is too fast for an instant-to-instant control, it must have been preprogrammed on the basis of acoustical cues received prior to its initiation (see also p. 145). Whether such a remarkable performance also forestalls truly protean escape, remains to be seen. Moreover, as cautioned by Webster (1967 b), the alternative hypothesis, that captures are guided by mere contact of insects with the flight membranes, cannot be dismissed. It should be added that both possibilities may exist side by side, and would, in effect, be adding to capture success.

The upsetting of the target mechanism by protean displays is thought to involve "confusion" of the predator. However, it is doubtful whether the phenomenon is identical with the "confusion" arising from schooling of prey animals, especially if one considers that with schools, different species of predators are affected in quite different ways (see p. 123).

"False Heads" and Other Directional Disguise. Some predators have been found to predict the direction of the progression of the prey even though it is stationary (p. 151). The predator is aiming its strike ahead of the prey, on the track that it will move. This sort of interception relies on the fact that the escape route is fixed in relation to the prey's body and that the initiation of the strike triggers an escape response. Strangely therefore, the prey's ignorance of the impending attack would make it fail. The rarity of interception with still prey, as documented so far, may be more apparent than real, as can be inferred from the occurrence of "false heads" and associated behavior patterns, that serve to hide the true course of potential progression of prey animals. The prey animals concerned fall prey to many predator species so that the protective attribute can be assumed to be tailored to more than one. A number of lycaenid butterflies, cicadas, and snakes mimic a head at their caudal end; as a rule, a "false head" does not mimic the particular head of its possessor. A false head may be a lure or deflective device to protect the true head from attack or to act in still other ways upon a

---

[12] In the first flash the bat's mouth is open although horseshoe bats (Rhinolophidae) normally keep the mouth closed and breathe and emit their orientation sounds through the nostrils.

predator (Wickler, 1968, p. 72 seq.). In the insects mentioned it is obviously to disguise the true route of escape from the attacker. The hind wings of some lycaenids bear white pseudopupils in a black field protruding sideways (Fig. 56). The elongated tips of opposite hind wings are crossed and terminally broadened so as to suggest two antennae with their terminal knobs. The true head and the true antennae are inconspicuous. In addition, four different behavior patterns tend to disguise the direction of taking off from a watching predator (Curio, 1965 c): (1) In several *Thecla* species the hind

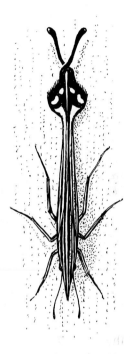

Fig. 56. 'False head' and head-down resting position of *Thecla linus* (Lycaenidae) on vertical substrate in Trinidad. From above. (From Wickler, 1968, p. 76)

wings are moved against each other so that the false antennae move up and down in alternation. The true antennae remain still. (2) When landing on a horizontal leaf *Thecla togarna* swings round by 180 degrees so that its head points to its previous flight direction; this maneuver happens so rapidly that it is hard for the human eye to see. After landing *Thecla* walks forward in the direction of its true head. (3) The related *Deudoryx* can, in addition, even walk in the direction supporting the mimicry of the false head. (4) Four genera with false heads as in Fig. 56 and two "normal-tailed" genera, rest head-down on a vertical substrate, which is unusual for butterflies. One of the latter species (*Thalicauda nyseus*) lands head-up and will only rest when it has turned around. As usual all species mentioned fly (true) head first when taking off. Thus a predator may be unaware of the new flight path not only by the false head but also by several movement patterns aiding in disguise. If this is accepted it also implies that a predator which has observed the landing of *T. togarna* extrapolates from the previous flight direction to a future one (Curio, 1965 c). This, of course, holds true only if walking in

155

between has not revealed the true front end and hence starting direction of the butterfly. It is, for instance, unknown in which way a predator would resolve the conflicting evidence of a false head and the normal forward walk true head on. These and related questions badly need rigorous testing with some of the potential predators. In addition to their possible role as an indicator of a wrong flight path, false heads of a number of lycaenids have been found to act as lure that direct a predator's pecks onto a non-vital part of the body (Mortensen, 1917).

V. Exhausting Dangerous Prey

There are prey animals that attempt to retaliate to attack by wounding or poisoning the predator. By its greater skill and stamina the latter will generally tip the balance in its own favor. Prolonged fights are restricted almost entirely to avian and mammalian predators. An adult moose surrounded by a pack of wolves is a dangerous quarry that can kick at its enemy with all four legs. Even when wounded a moose is often strong enough to stand its ground; its counter-attacks and the loss of blood eventually lead to a state of exhaustion so that the pack can bring it down with little risk (Mech, 1970, p. 215). By contrast, the pack will leave a healthy moose within minutes and hence must quickly recognize when a quarry is vulnerable (p. 129). Less powerful ungulates, e.g. wildbeest, have been reported to go into a state of shock when bitten by the pursuers so that there is no retaliation (Kruuk, 1972 b, p. 149). Solitary young pollock (*Pollachius virens*) or silversides (*Menidia menidia*) when pursued by piscivores have also been thought to enter a "state of shock" after desperate attempts to escape and thus fell easy victims (Radakov, 1973, p. 126). But what appears a state of shock in this case may well be related to "freezing", a common anti-predator defense; an operational definition would help to distinguish between the two behaviors.

In general, exhausting a belligerent prey takes a more active form than exercised by a pack of wolves. The predator lunges at the prey time and again and thus elicits retaliating strikes that finally bring about exhaustion of the prey so that the predator can deliver the deadly bite. A mongoose (*Herpestes i. ichneumon*) responds to a cobra (*Naja naja*) immediately by leaping from side to side. During these "mock attacks" it offers a continually changing and hence probably confusing target. Later on it repeatedly lunges at the cobra's head. The cobra in turn raises the head, with the "shield" erected, and lunges toward its tormentor. However, it can strike only down and forward as far as the raised forepart of the body allows. The mongoose evades the lunges by leaping up and sideways and tries to bite into the neck or elsewhere into the body from above. There are many sham attacks with no true biting intent (Fig. 57; see also Mendelssohn *et al.,* 1971; Ewer and Wemmer, 1974). Attack and counter-attack of both species occur almost simultaneously, which means that their response time must be about equal. Attacks occur in "rounds" of 3 – 6 sec. Bites that hit the mongoose subcutaneously are not necessarily fatal since it can tolerate the sixfold dose of what would kill a rabbit (Dücker in Grzimek, 1972, p. 165 seq., Vol. XII). Yet envenoming strikes from vipers (*V. xanthina*) make inexperienced young mongooses approach more and more carefully. In fact, attack may be discontinued altogether if the bites have hit the more sensitive areas in the vicinity of the eyes or the throat (Mendelssohn *et al.*, 1971).

Much more deliberately but with equal determination, a crowned eagle (*Stephanoaetus coronatus*) tired a cobra; while closing in, it parried the snake's lunges by throwing back its head each time (Meinertzhagen, 1959, p. 104).

Fig. 57 a – d. Mongoose (*Herpestes i. ichneumon*) fighting poisonous viper (*Vipera xanthina palaestinae*). Photographs (a) to (c) from 16 mm film taken at 64 frames/sec by A. Shoob. (a) Mongoose makes mock-attack, viper coiled ready to strike. (b) Mongoose quickly retreats while viper is still parading backwards. (c, d) Viper lunges at retreating attacker. Note unsharpness of pictures (b) and (c) due to rapid movement of both animals. (From Mendelssohn, *et al.*, 1971)

Quantitative investigations of such fights are virtually lacking. It is unknown, for instance, whether fatigue of the victim affects the quickness of its deterring counter-attacks or its accurracy of aiming. What makes the mongoose finally deliver the killing bite into the head/neck region remains an open question. The cat family gives a clue to the answer. Some of its members have been observed to overpower the quarry by exhausting it. From the description of a fight between a rat and an ocelot cat (*Leopardus tigrinus*) by Leyhausen (1973, p. 27) it becomes clear that it was sheer exhaustion induced by the defensive turning over on its back by the rat rather than deficient precision of its bites which made it finally succumb after 32 min.

Humphries and Driver (1970) have argued convincingly that several predators when attacking together are capable of confusing the prey. As they take turns in making incomplete attacks, the prey cannot counter-attack effectively since the direction of the attacks changes ("multiple lure display"). The erratic succession of attacking individuals is reminiscent of protean displays as a defense mechanism by prey animals. According to Meinertzhagen (1959, p. 41) ground hornbills thus tackle large snakes, including the extremely dangerous black mamba (*Dendroaspis polylepis*). From three to seven birds close in on the snake, walking sideways and using their stretched wings both as lure and shield. When the snake hits one of the trailing pinions their owner whirls around and pecks, then the next hornbill rushes in to peck and so on. Either by the many pecks or by a final concerted attack of all party members the snake is eventually killed (see also Axell, 1956; Young, 1963; Humphries and Driver, 1970). Similarly, on Galapagos one or two mockingbirds (*Nesomimus sp.*) together kill the giant scolopender (*Scolopendra galapagoensis*) by many furious pecks. The pecks are primarily directed at the head region and will sometimes lift the animal from the ground, only to release it again immediately. It tries to cling quickly to any molesting object and administer the poisonous bite (which to humans is extremely painful). When incapacitated the scolopender is cut into large chunks which are devoured or fed to nestlings (pers. obs.). Babblers (Timaliidae) appear to kill wasps in a similar way (Carpenter, 1942).

The most important feature of real and mock attacks in a prolonged fight is their quick advance and quick retreat. Predators that attack at high speed such as salticids, piscivores, or cats that leap onto their prey, are carried forward by their momentum so as to make a quick retreat unfeasible. Therefore mock attacks are the domain of predators that do not lose ground control during the assault or are expert fliers (see p. 149).

An elaboration of the "multiple lure display" has been reported for the dwarf mongoose (*Helogale undulata rufula*) by Rasa (1973). Here the same group member, the "strike instigator", undertakes to strike repeatedly at the head of a snake whereas other members, the "disorienters", deliver quick, distracting bites into the tail end. The α-male finally kills the exhausted reptile. It has not been ascertained how far the roles adopted by group members in one fight will be the same in another (see also Chap. 5. B.III. 3.).

VI. Insinuation

A hitherto unique tactic that inures the prey to the presence of the predator has been observed in a bug (*Nabis sp.*) by Arnold (1971). When trying to insert its proboscis into a caterpillar this latter performs vigorous defensive actions – which may actually injure a predaceous bug (see Morris, 1963). In spite of these the bug repeatedly probes the larva by placing only a fore foot on its body. Initially each contact elicits violent movements of the larva's thorax region that cause withdrawal. With repetition, however, the defensive movements subside until the bug can maintain foot-body contact with impunity. Then the proboscis is placed directly beside the foot and the prey is pierced as the foot is moved away. Thereafter any resumption of struggling by the larva usually abates quickly. It seems obvious that the defense of the larva breaks down as it habituates to repeated mechanical stimulation.

Essentially the same situation, though on a much more extended time scale, seems to exist in two birds. Laughing gulls (*Larus atricilla*) are the chief cause of egg mortality in royal tern colonies but are scarcely fended off by the terns. Buckley and Buckley (1972) suggested that the terns have become habituated to the continuing presence of the gulls with which they live associated all the year round, even utilizing the same winter quarters. All defensive responses are directed exclusively to occasional egg-robbers such as other gulls or crows (see also McNicholl, 1973). The detrimental effects of habituation to the royals seem to be mitigated by the gulls stopping to feed on them as soon as the chicks hatch, quite unlike their European counterparts, blackheaded gulls, which readily devour larid chicks once the opportunity arises. The spurning of royal chicks by the laughing gulls is so improbable that it gives the impression of "prudent predation" in a wide sense as a result of a continuing co-evolution of both species.

VII. Scavenging and Cleptoparasitism

1. Modes and Extent

Scavengers or carrion-eaters feed on carcasses or on scraps and offal left by other animals. The scavenging mode of life seems to have arisen in all major groups of free-moving animals. If scavengers appropriate prey from others or from hunters they may be called cleptoparasites (Rothschild and Clay, 1957, p. 10). As a rule, habitual cleptoparasitism occurs in animals that scavenge a great deal, i.e. in certain insects and birds (Fig. 58), while a number of mammalian carnivores scavenge or rob food merely as opportunity arises. Among wasps several genera in the Pompilidae (= Psammocharidae) and Sphecidae indulge in cleptoparasitism, or "cuckoo" reproduction (Wilson 1972, p. 377 seq.). To these may be added several families of bees where it is a common phenomenon (Halictidae, Anthophoridae, Megachilidae, Apidae). The cleptoparasite female digs her way into the nest of a female of one or more other host species. She then deposits her eggs onto the food with which the host has provisioned its progeny. The parasite then either eats the host egg (e.g. bembicine Sphecidae, some bees), or, more commonly, leaves the destruction of the egg to its own larva which is equipped for that end with sharp, falcate mandibles (e.g. Megachilidae). In addition, there may be physiological interaction between two or more parasitoids; it can be mediated by the production of specific inhibitors or by developmental arrest caused by competition for physiological resources (review Fisher, 1971). "Social parasitism" involves the intrusion of one parasitic female into the nest of another social species, usurping the position of the queen. The phenomenon and its extension through the social wasps and ants has been excellently reviewed by Wilson (1972).

All of the larger African carnivores down to jackals (*Canis* sp.) steal meat from others at one time or another. The cheetah is an exception, as it must be content with 40% of the weight of its total prey; or, to put it differently, it has to surrender 12% of all its own quarries to other carnivores. As a consequence, cheetah devour their quarry hastily, being constantly on guard

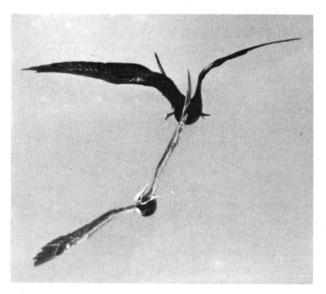

Fig. 58. Redfooted booby chased by frigatebird to make it disgorge food which will be snapped up by the tormentor. (Photo courtesy Rüppell)

against scavengers (Schaller, 1970, p. 38). This inferiority is offset by the cheetah's 70% hunting success which exceeds that of nearly all its food competitors (Schaller, 1972, p. 455).

Powerful scavengers such as spotted hyenas secure up to 26% of their carcasses by scavenging. They steal meat from seven predators, i.e. lion, cheetah, leopard, wild dog, jackal, vulture, and man. Furthermore, as active and extremely successful night hunters, they compete to a marked degree for the same prey animals as lion and man. In addition, hyenas scavenge during the day from the same sources as lion, leopard, jackal, and vulture (Kruuk, 1972 b). At their kill hyenas are dominated by lions which also scavenge and thus secure at least 16% of their food. Of all meat scavenged by lions, 42% is taken from hyenas (Schaller, 1972, p. 213 seq.).

Many birds force others to surrender their food by a variety of methods. For example, bald eagles (*Haliaëtus leucocephalus*) snatch fish from the talons of osprey. Professional pirates such as skuas (*Stercorarius* sp.) commonly chase and buffet their hosts in mid-air until food is released from the beak or disgorged. If the host is unable or unwilling to disgorge, it may even be killed as, for example, vultures by bald eagles (Meinertzhagen, 1959, p. 123). With much less effort parties of African thrushes (Turdidae) and bulbuls (Pycnonotidae) rob prey from driver ants, in addition to insects flushed by the foraging columns (Rothschild and Clay, 1957, p. 254).

As a rule it can be assumed that scavenging from others saves energy which would otherwise have to be invested in hunting, though some observations in birds are not easily reconciled with such a view. The pirating habit seems so deeply engrained in great skuas that these compel gulls (*Larus*) to surrender their food although the food would have been available much more easily as the gulls were picking up morsels in the wake of a ship in full view of the accompanying skuas (Meinertzhagen, 1959, p. 59). Similar behavior can be observed in the African fish eagle with a still greater discrepancy of energy investment between piracy and hunting (Brown, pers. comm.). These observations seem to suggest that to chase and pirate other birds for food is rewarding in itself, i.e. confers an advantage beyond the mere obtaining of food. This explanation recalls the possibility that predators switch prey for the sake of switching hunting methods rather than introducing variety into their diet (see 1. C. I.).

Many potential hosts clearly exhibit the booty they carry. A particular problem for avian pirates arises if only part of the host population brings in food, e.g. to feed the brood, after it has been swallowed and hence is no longer visible. But pirates obviously can pick laden birds and concentrate on them. The behavior of the host may well change as a consequence of carrying food and provide the pirate with the diagnostic cues (see p. 77). Caracaras (*Polyborus cheriway*) enhance the probability of robbing brown pelicans (*Pelecanus occidentalis*) by attacking solely the incoming ones; those leaving the colony never carry fish with them and thus never offer a reward (Meinertzhagen, 1959, p. 146). In many other cases it is not known whether or in which way the pirate identifies potential hosts correctly.

2. Cleptoparasitism and Competition

The importance of the occurrence of cleptoparasitism lies in that it is obvious evidence of competition for food. The competitive relationships between predator – scavengers have been regarded as "biological rank order" or "biological precedence" (Meinertzhagen, 1959, p. 27). It may be based on claw-and-fang actions in a Victorian sense but more often the outcome depends on subtle circumstances and thus by no means on a fixed "rank order" (Kruuk, 1972 b, p. 128 seq.). For example, wild dogs at their kill normally give way to hyenas which, in addition, are attracted by them even when no kill is present. Wild dogs rarely succeed in chasing off hyenas from their kill though ownership may change repeatedly. Adolescent hyenas sometimes manage to eat with the dogs from the same carcass although they are clearly inferior to the dogs. The latter only snap at them occasionally, apparently because the youngsters exhibit submissive behavior by crouching with their ears flat and their mouths half open, and show none of the customary fleeing or active defense of the adult. The competitive relationship between both species "may well depend on the interaction of a number of factors such as the number of each species, hunger state, previous experience, and individual differences" (Kruuk, 1972 b, p. 141). A similar state of affairs holds for the complicated relationship of mutual benefit and competition between hyenas on one hand and vultures, lions, and jackals on the other. The hyenas probably more often provide food than they take it.

The balance between competing scavenger – predators is sometimes so delicate that a carcass may change ownership up to four times, involving four different birds and a dog, with either no or only a slight scuffle taking place (Meinertzhagen, 1959, p. 29). The weaker often outwit the powerful ones, with the crow family, for example, stealing prey from much larger raptors, or wolves from a grizzly bear (*Ursus arctos horribilis*) (H. and J. van Lawick-Goodall, 1970, p. 109). Moreover, the weaker predator may effectively forestall robbery from the superior. Thus a glutton (*Gulo gulo*) drives puma and brown bear from its kill by using its stink glands (Herter in Grzimek, 1972, p. 67, vol. XII). Or African leopards (*Panthera pardus*) drag their quarry up into trees where it is safe from their numerous carnivore competitors. As a corroboration, Ceylon leopards eat their kill on the ground where they are not molested by competitors (Eisenberg, pers. comm.). Insects use pheromones widely to steal food from others (Hölldobler, 1973), or to repel others from portions of the same prey item (Fuldner and Wolf, 1971). The larva of the beetle *Aleochara curtula* feeds on the pupae of blowflies (*Calliphora erythrocephala*). After intruding into the puparium of its host, the larva seals the hole it has made with viscous feces. These contain a repellent that makes any later-coming larva avoid the sealed hole and drill another in a more rostral portion of the pupa. Thereby the parasites avoid too close contact within the host.

Forestalling robbery may even take the form of territoriality in that the robbed species are defended against other scavengers. Glaucous gulls (*Larus hyperboreus*) defend rafts of eiders (*Somateria mollissima*) against other gulls with the exception of immature conspecifics and great black-backs (*Larus marinus*). Larger rafts are controlled by more "owners" than smaller ones. The eiders are robbed of the shells (*Mytilus*) they expose. The idea of territoriality is strengthened by the fact that defensive actions are only seldom followed by immediate reward (Ingolfsson, 1969).

Competition may climax in killing of the inferior by the dominating species. These actions are often difficult to separate from mere killing for food. In fact, raptors kill others and owls (Meinertzhagen, 1959, p. 151), or vice versa (Craighead and Craighead, 1956, p. 208 seq.) and feed on them, and leopards kill and eat other cats (*Felis serval, F. caracal*) (Hendrichs and Hendrichs, 1971) and jackals (Schaller and Lowther, 1969). By contrast, lion kill but then abandon hyenas (H. and J. van Lawick-Goodall 1970, p. 184 seq.), wild dogs, other large cats, and man. Moreover, the facial expressions of lions when killing these animals are identical with those in intraspecific fights which are also dictated, at least in part, by competition. By contrast, when killing true prey a lion's face remains entirely calm (Schaller, 1972, p. 382 seq.). These facts favor the view that some killing is caused by a motivation that is different from hunger (see also 1. A. IV. 2.).

3. Counter-measures of the Robbed

The host in turn employs a number of counter-measures to forestall attack. These have been most carefully analyzed in cleptoparasitic interactions of puffins (*Fratercula arctica*) and arctic skuas (*Stercorarius parasiticus*) by Grant (1971). Incoming puffins flying towards their colonies are laden with several fish, sand-eels and fry of herring and gadids. The fish that are carried transversely in the bill are exposed to view and are the target of the skua's attack. To appreciate the defensive actions of the puffin a knowledge of the tactics by which the skuas try to maximize success will be helpful: (1) They chase puffins far from colonies where these are more likely to surrender food than close to their burrows. (2) Skuas position themselves on flat slopes close to burrows to intercept with a landing host. (3) They attack low-flying ones, possibly to attain greater speed in stooping from above. But this tactic is ambiguous because if the puffin is flying too low the food dropped may hit the ground before the skua is able to recover it; and on the ground *Larus* gulls and ravens dominate over the skuas which thus lose the booty to other scavengers. (4) Chases involving more than one skua seem to lead to food being dropped more frequently than those with just one skua. In groups of two or three pursuers, position in the chase is not an important determinant of individual success. The optimum number of pursuers will be determined by the opposing tendencies for probability of food release increasing with number of pursuers and by the probability of each skua securing food decreasing with increase in number of participants. This point and the evolutionary implications which follow from it will be dealt with more fully below (5. C. III. 3.).

Counter-measures of the puffins partially match the attack tactics of the skuas and include: (1) Incoming puffins turn back when spotting a skua on their way to the burrow and try to head in again. (2) Circling high if a skua is close to the burrow. Moreover, flying high forestalls attack as the skuas may not attain the speed necessary for attack if not flying higher than the victim. (3) Incoming puffins tend to aggregate in time (coefficient of dispersal 1.5 – 185.1) but depart dispersed at random on outward flights (coeff. of dispersal 0.87 – 1.30). Although clumps would tend to make the incoming puffins more conspicuous, skuas have difficulty in perceiving and choosing a host since clumps are made up largely of non-breeding birds without fish. Breeding birds take maximum advantage of the non-breeding birds by flying with them, using them as "decoys". For the same reason synchronized breeding may benefit puffins by allowing them to fly in clumps. (4) Some "psychological" factor makes puffins that are flying towards the burrow more reluctant to release fish than if chased far from it before delivery of the food to the chick. (5) Puffins that breed in a burrow on steep slopes can enter it more quickly than on flat slopes and are therefore less vulnerable to the skuas.

From all the stratagems listed above only the turn-back maneuver shown by the puffin is a direct response to the skua's presence which signals attack. All other tactics of both pirate and host may be regarded as forestalling measures of the opponent that may or may not happen. But if they do happen these tactics tend considerably to increase the likelihood of success of piracy and escape, respectively. Whether the tactics listed are especially designed to fulfil the function observed is not always clear. For example, nesting on steep slopes may have advantages aside from facilitating escape from skuas.

VIII. Tool-use

The use by an animal of an object extraneous to its bodily equipment allows an extended range of its movements or an increase in efficiency. Tool-using performances have been observed in animals as different as wasps, crabs, birds, and mammals. The most common occurrence is in food-procuring, agonistic, and defense contexts (review Hall, 1963). In one and the same species such as, for instance, the chimpanzee, there are certain behaviors and tools for food-procuring, e.g. angling for termites with sticks, and certain others for warding off larger carnivores (Kortlandt and Kooij, 1963). The frequency with which tool-use has been observed in free-living primates is rather low as compared to conditions of captivity. This has been taken as an indication that field records are still very incomplete. But as shown by Kortlandt (1966) species of primates differ tremendously in their propensity to use tools in captivity. Hence, the rarity of data for certain species seems real rather than an artifact.

The only experimental studies of the use of tools in the context of food-procuring have been conducted by Millikan and Bowman (1967) with the Galapagos woodpecker finch and by Jones and Kamil (1973) with blue jays. Apart from this, avian tool-using performance in connection with feeding behavior has been recorded anecdotally for Egyptian vulture (*Neophron percnopterus*) and the black-breasted buzzard kite (*Hamirostra melanosternum*) which hurl stones against birds' eggs (*Struthio camelus; Dromaius novaehollandiae,* Otididae) to break them (Chisholm, 1954 in Millikan and Bowman, 1967; H. and J. van Lawick-Goodall, 1966). The use of a three-inch leaf petiole to probe into a crack by one individual warbler-finch (*Certhidea olivacea*) (Hundley, 1963) seems to be no species-specific behavior.

The woodpecker finch probes with a stick or a cactus spine into cracks, holes, and under loose bark to stab, chase, or perceive by means of tactile stimuli prey animals hidden in those places (Bowman, 1961; Eibl-Eibesfeldt and Sielmann, 1962). A similar behavior has been recorded for the congeneric mangrove finch (*Cactospiza heliobates*), though for only one out of five

164

males (Curio and Kramer, 1964). The woodpecker finch picks up a stick somewhere, or breaks one off its substrate. If the stick is too long or forked, the woodpecker finch twists off and discards one prong, or shortens the stick and retains the shorter piece as a more manageable tool. Alternatively, too long a stick is seized in the middle, or the bird flies up, and while on the wing, successfully uses the stick to dislodge a bait mealworm from a slot (Millikan and Bowman, 1967). One out of six captive birds tried a tool made of three short sticks glued together at right angles, but it never succeeded in probing with it successfully. [The modification of a tool has been regarded by Napier (1963) as an advanced stage of tool-use, attained e.g. by chimpanzees.] In general, the view of a prey in a crack initiates tool-use as borne out by a comparison of baited and unbaited slots, especially if the prey cannot be reached with the beak. Quite often, however, one bird picks up a twig when no food is present. Possibly the twigs themselves initiate the behavior. A finch in the wild may carry a spine over several yards without having probed at the place where, after landing, it searches for prey (pers. obs.). This never happened in Bowman's captive birds.

When hungry, captive woodpecker finches were found to use more tools, as measured by the number of toothpicks removed, than when sated (Millikan and Bowman, 1967; see also Jones and Kamil, 1973). Furthermore, when a bird happened to hold a stick in its beak it could use it as a weapon with which to threaten a conspecific at the feeding table. Hence, apart from feeding motivation there may be others which cause tool-use.

There were many differences in detail among the six captive individuals. One male even failed entirely to probe with a twig although it cocked its head and held food under its feet like the other birds. Likewise only one out of five mangrove-finches followed closely by Curio and Kramer (1964) used tools. These observations suggest that individual learning plays a role in shaping the performance. A fledgling mangrove-finch that was still fed by its parents mandibulated with a stick without probing with it (Snow, pers. comm.). Possibly learning is by social facilitation (copying? See p. 79). A captive Darwin's finch (*Geospiza conirostris*) male lived for a year adjacent to a cage with two adult woodpecker finches which often passed sticks through to the individual *conirostris*. The latter, after some time, not only picked up twigs, but also stuck them in cracks. On Hood Island, the home of this bird, woodpecker finches are absent. Despite many attempts by Millikan and Bowman (1967) to teach tool-use to other species of Darwin's finches by caging them with woodpecker finches, no other bird ever showed any sign of tool-use. On the other hand, Jones and Kamil (1973) have evidence to assume that six of their caged blue jays "copied" from each other the habit of raking food pellets with pieces of paper which they could not obtain otherwise.

The ability of woodpecker finches to use sticks successfully in obtaining food indicates the presence of a special behavioral potential which is to be distinguished from other manipulative abilities such as string-pulling. Here, woodpecker finches proved to be no more proficient than species from the American mainland (Millikan and Bowman, 1967).

Table 8. Occurrence of mutilation of prey that generally does not lead to whole-sale consumption by attacker

Predator taxon	Prey	Body parts eaten

Aeolidiidae
  Aeolidia
  papillosa
	8 species of anemones	All except tentacles

Petromyzonidae	Numerous fish, few whales	Muscle tissue, blood (fish)

Characoidei [a]
  ≧ 10 species
  ≧ 5 genera
	Numerous fish	Scales, fins

Cichlidae [b]
  12 species
  6 genera
	Numerous fish	Scales, fins, eyes

Blenniidae 2 species [c]	Numerous fish	Fins (*Aspidontus taeniatus:* from rear end)
Blennius *galerita*	Barnacles (*Chthalamus stellatus*)	Filter apparatus when extruded [d]
Runula *rhinorhynchus*	Numerous fish	Skin pieces mainly around eyes
Trichomycteridae Many species	Numerous fish	Scales [e]; gill filaments, blood
Gobiidae 2 species	Sea urchins	Tube feet, pedicellariae [f]
Gobiesocidae 4 species [g]	Sea urchins, crinoids	Tube feet, pedicellariae; pinnulae of crinoid host

[a] Including fin-eating Ichthyboridae and *Serrasalmus* plus scale-eaters.
[b] 4 *Plecodus, Perissodus microlepis,* 2 *Corematodus, Haplochromis welcommei* eat scales only, the latter three species from caudal fin; *Genyochromis mento* fins in addition; *Docimodus johnstoni* fins; *Haplochromis compressiceps* the only eye-eater (Wickler, 1966; Fryer and Iles, 1972, p. 99).
[c] *Aspidontus taeniatus; Hemiemblemaria simulus* associates with its schooling model.

IX. Mutilation

A small number of predators have taken to the habit of mutilating their prey for food (Table 8). In part the habit is due to a low preference for the whole of a prey or to its defenses as in the anemone-eating nudibranch *Aeolidia papillosa;* attacked anemones roll off to the side, move away on their pedal disc, or repel the attacker by virtue of their nematocysts and/or their acontia.

Table 8. (Continued)

Remarks	Reference
Whole-sale consumption increases with prey preference	Waters (1973)
Some eat also carrion	Kühl in Grzimek (1970, p. 34 seq., vol. IV)
Other additional food in at least some species	Roberts (1972); Mathes (1961) quoted by Roberts (1972)
Two species (*Corematodus*) each mimic their victims with whom they associate. One (*Plecodus straelini*) exclusively scale-eating (in aquaria)	Fryer and Iles (1972, p. 85 seq.); Wickler (1966)
Mimic cleaning wrasses (*Labroides dimidiatus, Thalassoma bifasciatum*)	Wickler (1961, 1963, 1968, p. 159 seq.)
Opportunity for mutilation only within swells of surf zone	Soljan (1932)
Also feeds on invertebrates	Randall and Randall quoted by Wickler (1968, p. 175)
Bloodsuckers, temporarily beneath victim's operculum	Refs. in Roberts (1972)
Feed also on small crustaceans	Teytaud (1971)
Besides tube feet *Diademichthys deversor* takes other food as well	Refs. in Teytaud (1971)

[d] Gut found replete with filter apparatus and other body parts.
[e] *Apomatoceros alleni.*
[f] *Gobiosoma multifasciatum* pedicellariae only found so far; *Ginsburgellus novemlineatus* hides beneath host (*Echinometra lucunter*).
[g] *Tomicodon eos; Lepadichthys lineatus,* and *Dellichthys morelandi* live on their crinoid (*Lamprometra klunzingeri, Capillaster multiradiata*) and with their echinoid (*Evechinus chloroticus*) hosts, respectively.

Additionally, the nudibranch often ceases feeding on a particular anemone and lets it go when an abundance of a preferred species is available. Only the most preferred species (*Epiactis prolifera, Anthopleura elegantissima*) are eaten without cessation (Waters, 1973). For the most part, mutilators are so small, as compared to their victim, that for purely mechanical reasons the latter cannot be consumed wholesale. An exception is the eye-eating cichlid *Haplochromis compressiceps* from Lake Nyasa; with its elongate pointed

167

snout it pecks eyes from fish that are not smaller than the ones it eats entirely at other times. Curiously, in the aquarium it robbed eyes from cyprinid fishes yet none from fishes of three other families (Wickler, 1966).

Unlike most cleptoparasites, multilators are often morphologically and behaviorally highly specialized to cope with their prey. Fryer and Iles (1972) have provided the most comprehensive account of the manifold adaptations of the scale-eating cichlids of African lakes. The form of the mouth, the dentition of jaws and pharyngeal bones, and the protrusibility of the premaxilla all form a highly specialized syndrome that has evolved independently several times within the family notwithstanding its similarly independent evolution in the characoids.

The dentition of scale-eaters resembles that of species that scrape algal "Aufwuchs" from rocks and plants (periphyton) and has probably evolved from this stage. By and large, scale- and fin-eaters seem to exploit predominantly the same "host" fish fauna though sound field data on this point are lacking. Yet as revealed by close inspection of Table 8, differential exploitation of the same "basic food" does occur. For example, *Genyochromis mento* from Lake Nyasa prefers the scales and fins of *Labeo* species to other species. In one characoid species (*Ichthyborus besse*) the habit of fin-eating varies intraspecifically with the absence of other fin-eaters (Daget, 1967 in Roberts, 1972). Moreover, different portions of the "host" body are exploited by different scale-eaters among cichlids and blenniids.

It is difficult to prove that feeding by mutilation is the only predatory habit of a given species. Nevertheless, such specialization seems to occur. Thus in aquaria *Plecodus straelini* has fed on nothing else than on scales of other fish (Fryer and Iles, 1972), and some trichomycterid catfishes have been found gorged exclusively with gill tissue and blood (Roberts, 1972). The addiction to only one species of "host" has been reached solely, as it seems, in the goby-echinoderm and gobiesocid (clingfish)-echinoderm associations though even here there are exceptions. In these associations one species of mutilator has been found on only one sea urchin or one crinoid to the exclusion of other hosts. The mutilator (e.g. *Ginsburgellus*) lives beneath the test of an urchin (*Echinometra lucunter*) which thereby also provides, besides food, protection for the fish commensal. Others seem to live permanently on a crinoid, as for instance a clingfish (*Lepadichthys lineatus*). The virtually stationary life habits and the permanence of these fish – echinoderm associations make clear that mutilation grades imperceptibly into ectoparasitism. In fact, the sea lampreys (*Petromyzon marinus*) have long been epitomized as the only parasites among the chordates. Further, the trichomycterid catfishes seem to leave their host fish after being gorged with blood and burrow to rest in the bottom, thus not differing too much from a bloodsucking ectoparasite.

Many small fishes are known to seek protection among the spines of seaurchins when threatened, or all the time (Magnus, 1963). Virtually nothing is known about the stimuli by which the near-to-permanent association with a host is mediated. In one case the host is recognized by visual cues as preliminary experiments with a clingfish – echinoderm association have shown

(Dix, 1969 in Teytaud, 1971). From this relative loose use of shelter the fish – echinoderm associations may have developed. If, as in these associations, a mutilator should find its host individual no longer profitable and other host individuals were available, it could and in all likelihood would change to another. Other mutilators, however, are bound to one host individual for a long time or possibly, as with many parasites, for their whole life. The moth ear mite (*Myrmonyssus phalaenodectes*) that destroys the tympanic organ of its moth host can hardly change its vehicle. Here then the mutilator must have a vital interest not to damage its host up to a point when survival is jeopardized. To minimize being predated upon together with its host by a moth-eating bat, the ear mite behaves adaptively in a special way. It assures that only one ear becomes destroyed. If a moth is parasitized by one mite already any mite coming later associates with it, probably aided by a pheromone trail laid down by it. In this way one tympanic organ remains intact and can thus warn the moth of approaching bats (Treat, 1958).

B. The Diversity of Hunting Methods

I. Prey-specific Methods

Description. A number of predators assume different hunting tactics with different prey (Table 9). *Argiope argentata*, an orb-web spider, discriminates between lepidopterans and all other insect prey blundering into its web. It administers a long immobilizing bite to lepidopterans and afterwards wraps them in enswathing silk. All other prey, except for the smallest items that are only briefly bitten and then consumed, is wrapped first and only briefly bitten before the spider returns to its hub. Moreover, while the latter prey is immobilized by throwing long enswathing filaments over it from several sides, a moth when immobilized by the bite is rather deliberately wrapped before being carried to the hub for being eaten. Discrimination between both classes of prey probably implicates mechanoreception or olfaction (Robinson, 1969 b; see also Harwood, 1974). The differential capture response is nicely adapted to the biological needs, for the loose wing scales enable lepidopterans to escape from spiders' webs (Eisner *et al.*, 1964).

Hunting techniques differ by but one or few elements of the whole predatory sequence (Table 9, No. 8; Fig. 59). The pigmy owl kills mice by repeated bites into the snout whereas songbirds, its other staple food, are killed solely by a firm grip with the talons. Another difference concerns the site of ambush. This suggests that the readiness to hunt for either prey is different prior to any hunting episode, so that the difference is not just a consequence of momentary stimulation from the prey. There are still other indications that the motivational basis of preying upon mice differs in other respects also from that of preying on birds (Chap. 1. C. I.).

Some changes of technique involve orientation of approach toward the prey during attack (Table 9, Nos. 2 A, 3) or that of the incapacitating strike

169

Table 9. The prey-specificity of hunting/killing methods and their adaptedness. * Interpretation by Curio

No.	Predator	Prey	Hunting/killing method	Adaptive	Likely reason(s) for +	Reference
1	Spider (*Argiope argentata*)	non-lepidopterous insects	wrap, then bite	+		Robinson (1969 c)
		non-lepidopterous very small insects	bite			
		lepidopterans	bite, then wrap		scales facilitate escape	
2	Cuttlefish A)	prawn, crab	approach from behind	+	flee backward; (formidable chelae) [a]	Neill and Cullen (1974)
		fish	approach from front	+	flee forward	
	B)	crab	leap onto crab with all appendages; kill: within 9 sec	+?		Messenger (1968)
		prawn	plop with pair of tentacles; kill: bite into abdomen			
3	*Haplochromis compressiceps* (Cichlidae)	fish	approach from behind	+?		Wickler (1966)
		fish eyes	approach from aside	+	necessitated by victim's anatomy *	
4	*Eteirodipsas colubrina* (Boiginae)	frogs	seize and swallow	+?	rodents escape more successfully (?) *	Reinhard and Vogel, in Grzimek (1971, p. 417, vol. VI)
		rodents	seize, strangle and swallow			
5	Florida alligator (*Alligator mississippiensis*)	water snake cottonmouth moccasin (*Agkistrodon piscivorus*)	seize and swallow bite, toss and throw about, bite, etc.	+	avoidance of poisonous bites by cottonmouth	Neill (1971, p. 240)
6	*Lacerta sicula*	flies	stalk, seize wing, press down	+	flies' high powers of escape	Milinski (pers. comm.)
		crickets	approach deliberately, seize			
7	Lace monitor (*Varanus varius*)	mice	seize, forcefully masticate and swallow	+	powers of escape *	Horn (pers. comm.)
		birds' eggs	seize carefully, crush and	+	egg contents would	

8	Great grey shrike	frogs	seize, beat and rub on ground, masticate and swallow	+	removal of obnoxious skin secretions *	Thielcke (1956), Cade (1967), Ullrich (1971)
		insects	snaps up			
		mice	bite across neck and carry off with feet	+	avoiding being bitten	
		birds	seize with feet, bite across neck [b]	+?	powers of escape *	
9	Pigmy owl	mice	bite into snout and neck grip with feet	+?		Scherzinger (1970), see Fig. 59
		songbirds				
10	Red fox [c] (*Vulpes fulva*)	mice	pounces upon		minimizes audible friction at ground cover; (better vision of rabbit) *	Murie(1936) quoted by Kruuk, 1964, p. 44
		rabbits	chases after			
11	Polecat	rodents	bite neck and shake toss and roll about and bite; mock attacks		exhausting viper; minimizes poisoning removal of obnoxious skin secretns	Eibl-Eibesfeldt (1955, 1958 a)
		vipers		+		
		frogs	toss and roll about and bite			Gossow (1970)
12	Lion, Serengeti	larger ungulates	knock over, suffocate by biting into throat or muzzle, or through nape	+	avoidance of horns, impenetrable skin (wildebeest) and others little effort?	Schaller (1972, p. 264 seq.)
		smaller ungulates	knock over, bite through nape	+?		
	Lion, Kalahari	gemsbok (*Oryx gazella*)	leap onto back, bite in the haunches, break back with jerking motion upwards and kill with ventral strangle-hold or dorsal neck bite	+	avoidance of horns	Eloff (1964)

[a] Small crabs (20 – 80 mm carapace width) may be captured either way. Crab sometimes turned by the arms so that the chelae are hold outwards away from the body.

[b] Loggerhead shrikes direct their bites onto the front end of wooden dummies as typified by eyes and a neck-line indentation (S. Smith, 1973).

[c] When hunting in dense herbage also cats employ the 'pounce' (H. and J. van Lawick-Goodall, 1970; Leyhausen, 1973, p. 11).

or bite administered to the prey's most vulnerable body part (Table 9, Nos. 11, 12). The situation prevailing in the killing behavior of the Serengeti lion has been intentionally simplified; complications such as communal hunting etc. have been omitted so that in fact more than the two methods illustrated in Table 9 are possible. Larger prey are bitten across the nape (Fig. 60 a) or the throat ("ventral strangle hold"), with both orientations of the bite often applied in succession to the same quarry. Or larger prey are suffocated by covering the muzzle with the mouth (Fig. 60 b). Smaller prey are merely bitten through the body or whatever part the lion happens to grab. Cheetah, like lions, bite smaller prey animals across the nape or puncture the skull whereas they strangle larger prey by holding a sustained, ventral grip across the throat (Eaton, 1970; Schaller, 1972, p. 319). Suffoca-

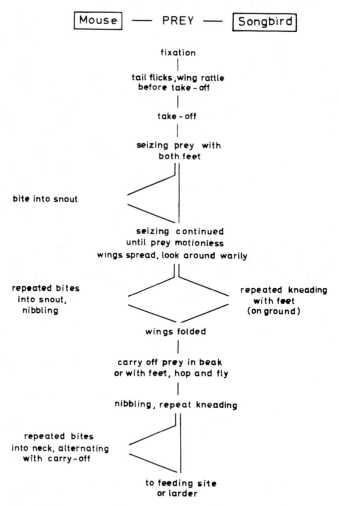

Fig. 59. Sequence of movements employed by pigmy owl in capture, killing, and handling of the two main prey animals. (After Scherzinger, 1970)

172

Fig. 60 a and b. Two killing techniques of lions with larger prey. (a) Lioness pulls down a zebra by clutching its neck and biting its nape. (b) Lioness suffocates a wildebeest by covering its muzzle with her mouth. (From Schaller, 1972; plate 32 (b) and unpublished original from author (a); plate by courtesy of the Univ. Chicago Press, © 1972 by Univ. Chicago)

tion and the "ventral strangle hold" are but two means by which lions avoid dangerous weapons. Kalahari lions overpower gemsbok despite the possession of formidable horns in a strikingly different way (Table 9, No. 12). In all three cases a difference in motor performance can be observed too.

The short account on the orientation of the killing bite touches on a problem to which much attention has been given. The subject is laden with controversies as regards, for instance, the killing bite of mammalian carnivores. This, and the complexity of the subject demand a treatment that would go beyond the scope of this book (but see below p. 174).

Other prey-related differences of capture/killing methods concern largely the motor performance of capture (Table 9, Nos. 1, 2 B, 6, 8, 10, 11) and/or killing (Table 9, Nos. 2 B, 4, 5, 7, 9, 11). Changes of predator group size and their significance will be treated below (5. C. III. 3.).

A given hunting behavior may serve several purposes. Polecats attack rodents by biting them across the neck and shaking them repeatedly. By contrast, to remove the obnoxious skin secretions from frogs (Gossow, 1970) or to prevent common vipers from biting, these animals are tossed and rolled about with the forepaws before they are killed by bites (Eibl-Eibesfeldt, 1955 b). In addition, the snake may at times be exhausted by repeated mock attacks which provoke lunges at the attacker (Eibl-Eibesfeldt, 1958 b).

In some cases the response differential is not absolute. Great grey shrikes capture songbirds generally with their feet, but on rare occasions strike at them with their powerful beak first before grasping them with their feet, i.e. employ the method usually used when killing rodents. The latter are killed invariably by bites across the neck (Cade, 1967). Furthermore, cuttlefish invariably "shoot" prawn with their specialized pair of tentacles and then bite them in the abdomen to dispose of them. By contrast, though crabs are killed the same way predominantly, they are also caught with the arms and killed within seconds, even without a bite (Table 9, No. 2 B). In 43 crabs tentacle attacks prevailed over arm attacks by 64% to 36%· The individual differences were considerable, perhaps as a result of past experience (Messenger, 1968). In this connection the fact deserves close attention that the arm attack on prey occurs later during ontogeny than the tentacle attack, which is present right after hatching. Newborn cuttles invariably shoot with their tentacles both prawn and crabs (Wells, 1958 quoted by Messenger, 1968). When growing up the tentacle response to crabs must at some time be dropped and is replaced by the arm response. The narrowing down of the range of eliciting stimuli of the tentacle response is also borne out by another finding. After amputation of both tentacles adults never attempt to capture prawn with their arms, but continue to catch crabs this way (Messenger, 1968). Hence, during adolescence any flexibility of the tentacle response disappears. An unusual prey situation, however, has been seen to evocate the tentacle response anew. A cuttle that had repeatedly failed to catch a crab dangling from a string above the water surface resorted successfully to the tentacle attack (v. Boletzky, 1972).

So far, differential responses to taxonomically different prey animals have been considered. It should be noted, in addition, that predators also subtly adjust their movements of seizure to the varying postures and orientations of their prey animals. For example, a red-tailed hawk (*Buteo jamaicensis*) strikes the prey [ground squirrels (*Spermophilus beechyei*), pigeons] with both feet simultaneously if it is oriented across or perpendicular to the line of attack. By contrast, if the prey is oriented parallel to the line of attack, the hawk places a lead foot on or just behind the prey's head 15 to 25 msec before placing its second foot farther back. The hawk keeps balance during this differential placement of the feet by appropriate wing compensations. Similarly, when attacking a pigeon in level flight a falcon (*Falco mexicanus,*

F. peregrinus) maintains a grip following contact, whereas during a "stoop" from a high vantage point it delivers a glancing strike with its talons which knocks the quarry to the ground (Goslow, 1971).

Adjustment also involves the strength of the attack or the bite. For example, herring larvae which typically s-bend in preparation to a strike (Fig. 53) bend stronger when about to strike at an evasive copepod than at a much more sluggish *Artemia* nauplius; they also then enhance the forward thrust by entirely unfolding their tail membrane (Rosenthal, 1969 b). Further, the amount of poison injected into prey by two species of poisonous snakes has been found to vary with prey size. Results depend on the species of snake studied. Whilst the amount of poison varied with the size of the prey rodent in one species (Gennaro *et al.*, 1961) it did so only in some individuals of the other (Allon and Kochva, 1974).

Teleonomic Considerations. Adaptedness of either of the alternate hunting methods implies that their employment would be less successful if directed against that type of prey which generally releases the alternate response. Although the problem of hunting success will be treated more fully below (Chap. 5. C.) a few remarks will be appropriate here. For none of the cases listed in Table 9 is information at hand that would permit relating hunting success to hunting method. Lion hunting methods differ strikingly in success rate though not recognizable with prey species (Schaller, 1972, p. 254). Therefore all suggestions as to the adaptedness of the respective hunting methods in relation to type of prey remain at present more or less plausible guesses. The problem appears particularly pertinent where more than two hunting methods have been reported as for example in the lace monitor (Table 9, No. 7), in certain herons with up to four (Kushlan, 1973), or in lions with up to eight methods (see Table 16 below). In Table 9 a hunting method has been provisionally designated + if, by its teleonomic consideration, it seems to be adapted to the prey to which it had been applied; only then should it be onerously called a "strategy" (p. 53). Hunting methods have been jointly designated + if it is the difference which seems to be adaptive, i.e. where at least one of the alternatives seems functionally "self-evident". However, what appears to be "self-evident" is often more complex than prima-facie evidence suggests. For example, fish, a most elusive prey for predators, are either approached from in front by one predator (Table 9, No. 2 A) or from behind by another (Table 9, No. 3), and a functionally plausible rationale could easily be construed for both techniques. Cases such as these caution against any interpretation without closer study of the teleonomy.

Some suggestive evidence as to the usefulness of a hunting behavior can be gained from its correlation with the type of the prey across predator species. For example, both lace monitor and polecat deal with frogs in a similar and most conspicuous way. The monitor rubs and beats a frog against the ground while firmly holding it with the snout. The polecat rubs a frog with the forepaws over the ground while administering interspersed bites across the back (Table 9, Nos. 7, 11). It is reasonable to assume that by the rubbing actions obnoxious skin secretions are possibly removed from the prey. The occurrence of functionally similar hunting behavior in two unrelated predators suggests that it has evolved independently in response to the same group of prey [13].

One would be inclined to think that specialists employing but one method of attack would have the advantage over generalists with a number of alternate tactics: As the number of hunting methods increases each one may lose precision; the argument rests, as should be stressed, more on intuition than on real data. Moreover, the choice of the appropriate method from a number of alternatives may become subject to error, as suggested by the

Table 10. Relationship between diversity of prey and hunting success of predators of black-headed gulls as reflected by the gulls' success in diverting attacks. (Modified from Kruuk, 1964, p. 114)

Species of predator	Threat towards	Over-riding tendency of defense behavior	Success of predators
Peregrine	adults	fleeing (formation flight; few attacks)	low
Harriers (*Circus*), red fox, man, stoat	adults and brood	ambivalent (attacks with gulls staying at a distance)	great
Crow, herring gull, hedgehog, oystercatcher, coot (*Fulica atra*)	brood	aggression (effective attacks close to the nest)	low

ambivalence of capture tactics of shrikes and cuttlefish (Table 9, Nos. 2, 8) described above. However, Kruuk (1964, p. 113 seq.), in an analysis of predation in colonies of black-headed gulls, found evidence that possible disadvantages of being a "jack-of-all-trades" may be offset by factors residing in the defence system of the prey. Those predators such as harriers, red fox, man, and to a lesser degree stoat, that took adults and their brood (eggs, chicks) levied the greatest toll (Table 10). The counter-measures of the gulls proved to be least effective [although there is no evidence for the harriers (*Circus*) to support this notion] because of vigorous direct reactions to their attacks (stoat) or because of strong fear as reflected in the gulls by staying far away from the interloper (fox, man). By contrast, the predators threatening the brood solely were either far less or not at all dangerous for the adults and could therefore be attacked more effectively. This in turn led to com-

---

[13] It is still unknown exactly which prey group has put the selective premium upon the rubbing response. Ranids are eaten more often by birds than are bufonids, yet some *Aquila pomarina* young reject frogs entirely (Kabisch and Belter, 1968 and Refs.). In accordance with their greater vulnerability *Rana* species possess, as a rule, distinctly fewer pharmacologically active and partly poisonous substances in their skin secretions than the *Bufo* species (Daly and Witkop, 1971, p. 502 seq.).

176

paratively light toll from the broods; although the hedgehog was most successful in breaking through the defence of the adults by displaying some degree of stubbornness. Predatory success was again low for peregrines, presumably because the gulls in their formation flights did not have to compromise between actions protecting themselves and others protecting the brood. In conclusion, the most versatile predators were the most successful because of inadequacies of the defence system of their prey animal rather than because of the superiority of their hunting behavior.

The inadequacies in defense result from the versatility of certain preddators. Those that threaten both adults and the brood may be regarded as two "functional enemies" from the gulls' point of view. Perhaps the gulls have difficulty, apart from those relating to especially frightening predators mentioned above, in choosing the right defense behavior to cope most effectively with each "functional enemy". If this is correct it would support the notion propounded above that a possible drawback of versatility of hunting is rooted in difficulties of choice of the most appropriate hunting method from a number of alternate ones.

II. Situation-specific Methods

Apart from the type of prey, constraints associated with the hunting situation as a whole need to be considered by the predator. These range from obvious physical constraints (Table 11, Nos. 1, 2) to prospects of the hunt in terms of the likelihood of escape of the targeted prey (Table 11, Nos. 3 to 5). Open water hunters like piscivores (Table 11, No. 1) and cuttles (Messenger, 1968) are carried by the momentum of their attacks beyond the point of prey seizure. The same holds for avian raptors when binding their prey during stoop in mid-air. By contrast, seizure of prey living on a substrate demands ability to stop suddenly as depicted above for octopus when catching a crab (Fig. 51). A slowed-down approach prevents the predator from injury in the impact of its attack. It also permits close inspection and dislodging sessile or slow prey; this may be the chief reason for the rapid forward braking movement of the large-mouth bass in this situation (Table 11, No. 1).

The factor(s) that determines which of several alternate routes is taken in a hunt are often not obvious, since a different hunting situation may not only affect the predator but also the behavior of the prey. For example, in the case of a red fox hunting for black-headed gulls (Table 11, No. 4) it remained unclear "whether these different techniques were a more direct reaction to the gulls' behaviour or whether the fox did actually take these environmental circumstances into account" (Kruuk, 1964, p. 48). A similar consideration applies to the different tactics of the cat (Table 11, No. 5).

At least some vertebrate predators sense acutely the moment a desired prey is going to disappear and thus give up hunting as a failure. Accordingly, predators attempt to time their attack so as to forestall escape. Sparrowe (1972) has analyzed what factors determined whether captive American kestrels attacked a (baited dummy) mouse (Table 11, No. 3). He found that

Table 11. The specificity of hunting methods as a function of the hunting situation.* Interpretation by Curio (for further details see text)

No.	Predator	Hunting situation	Hunting method	Adaptive?	Likely reason(s) for +	Reference
1	Large-mouth bass	water column on bottom	approach fast approach slow, stops	+?	a) Momentum of attack would prohibit dislodging sessile prey; b) Prey in mid-water (fish) more elusive *	Nyberg (1971)
2	Newts (*Triturus* sp.)	in water on land	snap with jaws strike with tongue	+	Resistance met by tongue in water and loss of stickiness *	Halliday (pers. comm.)
3	American kestrel	mouse in long runway mouse in short runway	attacks more often attacks less often but with shorter latency	+	Prospects of successful escape *	Sparrowe (1972)
4	Red fox	clear night and/or gulls on open beach overcast night and/or gull on nest	approach very fast stalk	+	Behavior of prey or their prospects of successful escape (see Chap. 5. A. II. 1.)	Kruuk (1964, p. 47 seq.)
5	Domestic cat	mouse in the open mouse close to cover	stalk approach fast	+	ibid.	Leyhausen (1973, p. 102)

the duration of exposure along a runway of 120 cm (Fig. 61) at a constant prey speed of 62 cm/sec crucially influenced the number of capture attempts and the rate of capture success. Contrast of prey against a jet-black background and percentage of cover, i.e. occluding black strips of paper, were less important. With a shorter runway subjects attacked less often; however, if they did, they made their decision to attack much more rapidly, thus increasing their chances of success. From this it follows that the raptors assessed the prospects of the hunt from the prospects of future prey exposure and hence from the total situation. This performance must relate velocity of prey to distance from the cover it aims at. Accordingly, under conditions of a good view of the prey, the duration of current stimulation by the prey cannot have played a crucial role. In a similarly complex way puma assess the circumstances of a hunt on a herd of deer (Hornocker, 1970). The way in which some of the large carnivores intercept the path of their prey (5. A. IV. 2.) also bears witness to an anticipation of the future whereabouts of the prey.

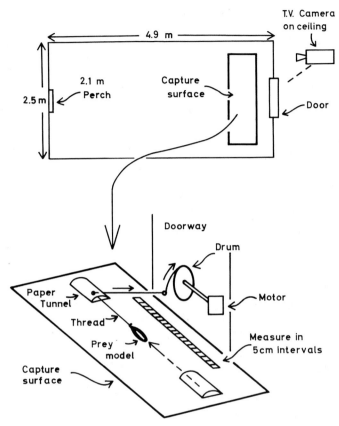

Fig. 61. Experimental set-up to analyze some of the factors which determine the decision to attack a moving prey model by an American kestrel. Before appearance of the prey from one of the paper tunnels the kestrel perched opposite and above the capture surface. (Redrawn from Sparrowe, 1972, courtesy of the author and J. Wildlife Management)

In conclusion, if escape chances are particularly good, approach during attack is either very rapid (Table 11, Nos. 4, 5), or the decision for attack is either abandoned or made especially quickly (Table 11, No. 3); this involves clear visibility and hence a good control of the environment by vigilance and little cover to support stalk. This can be interpreted to mean that the predator chooses a behavior that maximizes the likelihood for success although the prospects are poor. In fact, both ways combined, i.e. a very fast approach and a quick decision to attack, appear most appropriate to forestall escape of prey when the prey is likely to succeed in escaping. Conversely, if escape

179

chances are poor (e.g. cover far away, little control of the environment by vigilance) the predator should scrutinize the hunting situation more carefully. It should be noted that a fast approach may have the disadvantage of triggering flight, as has been found in a prey fish, from a longer distance than the normal one (Dill, 1973).

It will be remembered that especially quick or jerky movements of the prey also trigger a rapid rather than a stealthy approach in highly diverse predators. If such prey movements are indicative of high escape powers then again one would have to conclude that predators take a chance by rapid approach when the prospects of the attack are not particularly good (Chap. 5. A. II. 1).

III. Mechanisms and Causes of Predatory Versatility

1. General

On account of the relations between hunting/killing tactics and the type of prey (Table 9) predatory versatility might be predicted from variation of diet. An euryphagous predator is more likely to exhibit several tactics than a monophagous. The prediction, however, is a vague one. First, it does not allow for variation of tactics imposed by the boundless number of possible postures and orientations of even a single prey type (p. 174). Hence, a monophagous predator also needs some degree of versatility. Second, predators capable of only one predatory performance succeed in catching a broad variety of prey though in this case diet tends to vary predominantly in terms of species with similar anti-predator mechanisms.

Basically, two explanations have been suggested to account for the variation of feeding methods within a species (Fig 62). Either each individual (S, S' etc.) is specialized on few tactics (A), or is a "generalist" (G, G' etc.) capable of most or all tactics used by the species (B). In both cases it is assumed that each individual invests about the same effort as measured by hunting time, and second, that extreme activities leading to prey acquisition are performed less often than intermediate ones. The sum total of all individual curves is assumed to be the same for both models. Various combinations of both models seem to be feasible.

It must be considered now what sort of observational evidence would allow to distinguish between the two models. A sufficient condition for the identification of either model can be obtained if one considers the ranges of movements occupied by individuals as the fraction of what the species as a whole is capable of performing. In model A of Fig. 62 this fraction is rather small as compared to model B where all individuals share a large number of movements. In A only the most similar share part of these. This then yields a diagnostic feature by which to distinguish specialists from generalists: if one finds individuals whose ranges of tactics are mutually exclusive a case for model A would have been made. In each case the number of individuals to be scrutinized could, of course, depend on the amount of overall variation, width of individual range and other factors (for a mathematical treatment see Roughgarden, 1974). These considerations apply equally well to euryphagous predators with but one hunting pattern, for individuals may prefer certain types of prey over others; hence variation of perceptual mechanisms underlying prey recognition and prey selection also needs to be sought.

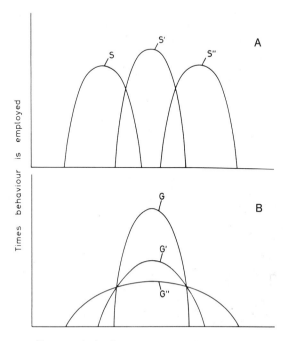

Fig. 62 A and B. Two descriptive models accounting for predatory versatility within a species. (A) The species is composed of numerous specialists *S, S'* etc., capable of performing but a very limited number of search, capture, and/or killing tactics. (B) The species is made up of 'generalists' *G, G'* etc., capable of a whole range of tactics typical for the species with some individual variation of range of performance. (Redrawn from Klopfer, 1962, Behavioral aspects of Ecology, by courtesy of Prentice-Hall Intern. Inc., Englewood Cliffs, New Jersey, USA.)

2. Individual Predatory Repertories

Dietary differences among individuals have been established for fish (Ivlev, 1961; Bryan and Larkin, 1972), salamanders (*Ambystoma gracile*) (Henderson, 1973), birds (Kruuk, 1964, p. 11; Harris, 1965; Shaffer, 1971; Dinsmore, 1972 and Refs.), and mammalian carnivores (see below). As a rule, individuals do not occupy exclusive ranges of the food spectrum, but rather differ from each other by preferring certain prey items over others. Dietary differences may reflect corresponding variation of perceptual mechanisms (1), searching (2), feeding behavior (3) and feeding rhythm (4).

181

(1) Discontinuous variation of innate prey recognition has been recently discovered in two reptile predators. While two variants of a snake exhibit a preference for either of two types of prey, variation in a Jamaican lizard (*Anolis lineatopus*) studied in our laboratory (v. Brockhusen and Curio, 1975) shows a unique rigidity of differences between individuals. Seventy five juvenile were kept from hatching in visual isolation from each other. When still naive for each of five about equal-sized (in terms of body length) prey animals, they were offered mealworms, cornbeetle (*Alphitobius diaperinus*) larvae, waxmoth larvae (*Galleria mellonella, Achroea grisella*), crickets, and woodlice (*Armadillidium* sp.) in succession. Individuals differed strikingly from each other in the acceptance of prey items. On first presentation of each of the five prey items individuals exhibited seven out of 32 possible combinations of rejection/acceptance (Fig. 63) which were due neither to hunger

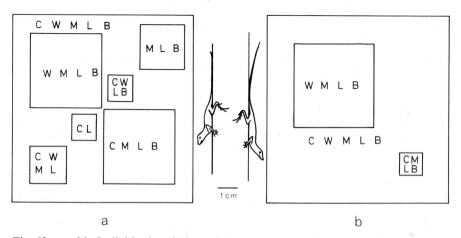

a   b

Fig. 63 a and b. Individual variation of the acceptance of five prey types on first encountering them in hatchling *Anolis lineatopus* (a) and after five weeks of feeding (b). Sizes of squares reflect the relative proportions of each behavioral type, represented by the combination of the items it had accepted; sketches of the animals at each respective age. Acceptance: *C* cricket, *B* cornbeetle larva, *M* mealworm, *L* waxmoth larva, *W* woodlouse. (Modified from v. Brockhusen and Curio, 1975)

nor to previous feeding experience with the particular item under examination; previous other items might, however, well have influenced the response. Waxmouth larvae were the only items that were eaten by all young. To this degree the species can be regarded as composed of "generalists". After three to four presentations of each item only three of these combinations were left, the majority (79%) of the anoles then accepted all five prey items, whereas two smaller percentages rejected either of two prey items; while 19% ignored crickets selectively, 2% did so with woodlice. It is important to note that "cricket-rejectors" are not later identical with "woodlouse-rejectors" (Fig. 63 *b*). Both types of "rejectors" ate the other four items as readily as the indiscriminate "acceptors". The discreteness of the two groups

182

of "rejectors" supports the view that the causes for the rejection of crickets and woodlice are different. This idea receives support from still another piece of evidence. "Rejectors" prove differentially liable finally to accept the previously rejected prey. Crickets are ignored for significantly more presentations than are woodlice.

Individual young differ strikingly in their responses to the very first prey item they see. Therefore individual differences exist independently of any information contained in the prey animals themselves. However, individuals are labile inasmuch as part of both groups of rejectors became wholesale acceptors. It remains to be seen what factors contribute to this change of preferences. In the two snake variants studied by Burghardt (1975) individual preferences are readily reversed by feeding either morph the alternate, little-liked prey. Alteration of prey preferences in the anoles must proceed in another way since preferences are not reversed but broadened. Whether this too is a consequence of feeding on the alternate, readily accepted prey, remains an open question.

The situation in both cases might be a "polymorphism" because variation is discontinuous. It should be borne in mind, however, that discontinuity is judged in terms of prey species and their acceptance/rejection. Whether discontinuity also prevails at the level of visual stimulus evaluation is another and unanswered question. Perhaps it could be tackled by bridging the gap between a rejected and one of the accepted prey species by means of offering intermediate prey items. If the proportion of the respective rejectors could be made to vary continuously with the acceptance of the prey species, individual variation of "intrinsic" sensori-motor machinery could hardly be regarded as discontinuous. If, on the other hand, the proportion of rejectors exhibited a hiatus somewhere along the array of intermediate prey items, a case for discontinuous "intrinsic" variation could be made.

(2) Some great tits, but not others, have been seen to detect caterpillars by peering upwards through the foliage instead of down- and sidewards (L. Tinbergen, 1960). It is unknown to what extent this influenced prey taken.

(3) Numerous examples most of which bear witness of (1) also illustrate differences of motor performance or of the propensity to employ certain predatory behavior:

Three individual green herons invented baiting and catching fishes (p. 144; see also 5. A. VIII.).
Several white pelicans (*Pelecanus onocrotalus*) in Giza Zoo (Cairo) used to drown teal that landed nearby by holding them with their bill under water and then devouring them (Meinertzhagen, 1959, p. 48).
Apart from mere knowledge of dietary differences between raptor individuals (e.g. Sulkava, 1964), various species have been observed to vary individually and to employ successfully tactics common to other species (Cade, 1960; White, 1962; Mead, 1963). A unique maneuver was used by a trained goshawk which used to catch red squirrels, a notoriously difficult prey, with great success: it alighted in the tree above the quarry so that it could follow the spiral-shaped escape path around the trunk and in its stoop grasp the squirrel with one foot (Saar, pers. comm.).
Chimpanzees vary in their propensity to use grass blades or sticks to probe for termites in their mounds. Hurling of branches against enemies like leopards varies

much more with the individual and is also less accurate (J. van Lawick-Goodall, 1968; Kortlandt and Kooij, 1963).

If communal hunters tackle a quarry jointly some of them may assume a role different from the rest of the party (see also 5. A. V.; 5. C. III.):

Amongst a pack of wild dogs or a clan of spotted hyenas it was often the same animal that killed the chased quarry (Estes and Goddard, 1967; but see Kühme, 1965; Kruuk, 1972 b, p. 116).

Pair hunts of various birds of prey suggest co-operation among partners with one being the "beater" and the other the hunter (see p. 203).

From a study of 18 wolf cubs from four litters Fox (1972) concluded that individuals differ markedly in their exploratory and fearful behavior when confronted with novel and/or moving objects. Significantly, the most exploratory individuals were the highest ranking in a litter and potential alpha animals for a forming pack. Conversely, the least exploratory individuals exhibited the least predatory proclivity and the most proximity tolerance and were hence the subdominant members of a litter. Although the genetic causation of these differences is distinctly unproven, Fox (1972) speculated that they form the basis of pack integration: a litter could not develop into a harmonious pack if entirely composed of "alpha" individuals; and it would not survive if it consisted solely of mute "non-killers". Significantly, Fox found canids which live solitarily and hence do not need packs to be behaviorally much more homogeneous within litters.

Henderson (1973) found dietary differences strikingly paralleled foraging stations (see also Bryan and Larkin, 1972) in larvae of salamanders in Marion Lake (Canada). Those that fed in the water column ate approx. 100 times more of a cladoceran (*Sida crystallina*) than benthic feeders which in turn fed more on other invertebrates, mainly amphipods, chironomids, odonates, molluscs, and leeches.

Specialization could be induced by environmental conditions. Freshly hatched and two-year old larvae were either kept with *Sida* or with sediment from the lake for one month each. When given a choice between *Sida* or benthic prey, larvae preferred the prey on which they had been reared

Table 12. Mean number of prey eaten per second year larva of *Ambystoma gracile* when given a choice between benthic and water column prey. Prior to that larvae were maintained for one month on only one of both prey. (Modified after Henderson, 1973)

Site of prey	Prey	Water column feeders N = 38	Benthic feeders N = 32
Water column	*Sida*	102.7	58.1
Sediment	Amphipods	6.8	13.6
Sediment	Chironomids	2.2	2.9
Sediment	*Pisidium*	2.5	3.1

(Table 12). The means for number of *Sida* or amphipods eaten by second-year larvae were significantly different (P < .05).

We cannot yet decide in what preferences salamanders differ primarily. However, benthic feeders consume a small portion of *Sida*, and, conversely, water column feeders a small number of benthic prey. These facts suggest that it is the foraging station which brings about the marked dietary differences. Prey and feeding site preferences are thought to be stable traits because (1) individual larvae were seen foraging either in the water column or on the benthos in successive days (Neish in Henderson, 1973), and (2) water column feeders were found to be consistently larger than benthic feeders (4.3 ± 0.3 cm vs 3.4 ± 0.2 cm snout-vent length). It remains to be explored whether water column feeders are large as a consequence of greater food intake, or whether they feed in open water because they are large.

(4) Individual domestic cats have been observed to hunt at different times of day (p. 37).

3. The Persistence of Individual Traits

Short-term specialization is borne out by various forms of object- or area-concentrated search for food (2. C., 2. B.). A more persistent or even life-long predatory specialization has rarely been demonstrated. Bryan and Larkin (1972) have shown that rainbow trout maintain individual dietary differences for up to six months, and there is similar though less good evidence for aquatic salamanders (see above). Trout specialization proved to be strongest over a few days and, surprisingly, approached random ingestion of prey items immediately thereafter in the majority of animals.

Another hint comes from populations that consistently exhibit special feeding traits not shown by others. Here, specialization of habit must have survived individual specialists, and since the older group members are the teachers of the younger they would appear to fulfil the condition of life-long specialization. Population differences of predatory habits are indicative of this sort of tradition:

Norway rats are known to feed on a large variety of plant and animal matter. The rat population on the island of Norderoog (West Germany) developed the unusual trait of killing birds after an outwitting stalk (Steiniger, 1950). Birds were presumably an enrichment of diet rather than the sole source of food. It is not known how this noteworthy habit spread in the Norderoog population. In the laboratory nursling rat pups have been shown to become conditioned to gustatory cues reflecting the flavor of the mother's diet by drinking her milk. These cues, moreover, are sufficient to influence dietary preference at weaning but an attraction to the site of feeding by the adults also plays a role (Galef and Sherry, 1973).

In the spotted hyena, clans of the same large population differ strikingly by their prey preferences (Kruuk 1972 b, p. 118). While two clans elected to hunt wildebeest taking 55 and 81% by species respectively, another clan specialized on zebra by 73%. This disparity was also found in the numbers of wildebeest and zebra actually eaten by these clans and so the difference cannot be explained merely by a difference in hunting success, in which, incidentally, the clans also seemed to differ. It can be safely ruled out that clan differences merely reflect differences in food availability in the clan ranges (1. C. I.). Specialization was not absolute, as switching from one to the other alternate prey occurs in all the clans studied.

Specializations reported thus far appear to conform to model *B* (Fig. 62) in that all individuals seem to share a basic repertory but deviate from others by special types of search, capture technique, and foraging site; as a result of the latter, marked dietary differences have been shown to arise in a salamander. In no case was there mutual exclusion between individuals as postulated by model *A*. The foraging site differential of the salamanders is suggestive evidence but some generalists seem to occur alongside both types of specialist (Henderson, 1973). It will now be examined whether mutually exclusive specializations do exist and how persistent they are.

Oystercatchers feed on a variety of estuarine animals such as crabs, clams, and worms. As demonstrated by Norton-Griffiths (1968) specialization of diet and of killing performance does occur. The near-to-exclusive preference for crabs rather than the usual mussels in one whole colony is evidence for the first sort of specialization. The second is unique as individuals are either "stabbers" or "hammerers" with respect to how they kill prey. In addition to "stabbing" or "hammering" a variety of other movements are used by all individuals, such as "levering" (Fig. 64) with the bill in an opened mussel; "prising" apart movable parts of a (killed) prey or of mud when probing; and "chiseling" off meat from the shell of a killed mussel or crab. In "stabbing" the bird directs a quick downward lunge with the sharp beak through the gaping posterior valve margins of a clam; the beak severs the strong posterior adductor muscle and thus disables the mussel. In "hammering" a series of hard blows are aimed at the most vulnerable point of the prey (mussel, crab), to kill it.

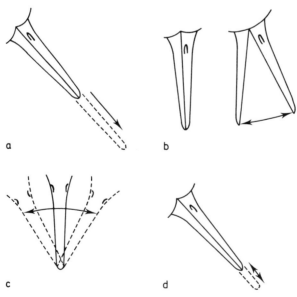

Fig. 64 a – d. Movements of the oystercatcher when killing and handling prey. (a) "Stab", differs from "hammer" by frequency and other details, e.g. orientation; (b) "Prise"; (c) "Lever", can be super-imposed upon a pivoting of the beak around its long axis; (d) "Chisel". (Adapted from Norton-Griffiths, 1968)

186

The skill of stabbing and of hammering is acquired from the parents, and so are certain prey preferences. This has been substantiated by cross-foster experiments. Significantly, both parents of a pair are either stabbers or hammerers. This concordance ensures an optimal transmission of skills but there are other functional and causal reasons as well. All young have the potential to develop either skill, which takes almost a year. Obviously, in about 20% of all birds this flexibility is retained as evidenced by the acquisition of a new killing technique in adulthood. A new technique must be acquired from adults other than the parents.

In conclusion, the life-long dichotomic specialization of skills in the oystercatcher means, for the majority of the species members, a realization of model A, i.e. of specialists with mutually exclusive ranges of predatory skills. An inclusion of the 20% minority which possess two killing techniques instead of only one, would conceptually result in a mixture of models A and B: generalists would exist along with two types of specialists (Fig. 65).

The stabber – hammerer dichotomy teaches yet another lesson. It was surmised above (p. 176) that as the number of hunting techniques increases, the effectiveness of every single one might deteriorate for reasons related to the accuracy of choice of the most appropriate pattern. It now appears that effectiveness might also suffer from the difficulty to perform two skills with equal precision, as is suggested by the small portion of "stabber plus hammerers"; moreover, the period to reach top performance is extremely long. The acquisition of either skill takes up to 43 weeks whereas worm-eaters fend for themselves at six to seven weeks; they only have to develop the simple probing for worms. It takes probably three to four years for a stabber or hammerer to become proficient in procuring prey, as (1) the body weight continually increases over this time, and (2) the adults do not breed before

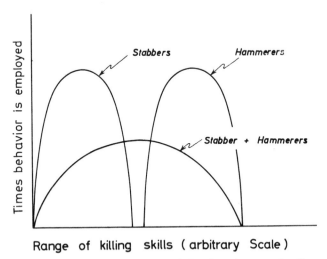

Fig. 65. Diagrammatic representation of individual specialization for two feeding skills in the oystercatcher, combining two models, (Fig. 62 A, B), into one with altogether three behavioral phenotypes. Based on the results of Norton-Griffiths (1968)

that age, i.e. seem to defer maturity until they are able to secure enough food for a whole family successfully. (It is unknown whether this argument also applies to the easy-going worm-eaters; they were permitted to breed sooner).

4. Predatory Specialization and Structural Modification

As is well known the castes of ants and termites, e.g. workers and soldiers, fulfil vastly different tasks within their society. Beyond this crude division of labor there exists differentiation of roles within e.g. the worker caste which is most pronounced in many groups of ants. Upon this sort of "caste poly-ethism" an "age-polyethism" may be superimposed in which workers go through a number of tasks the sequence and duration of which are species-specific; the program is, however, readily modifiable by external factors and individual behavioral ontogenies vary greatly in both content and timing (review Wilson, 1972, p. 163 seq.).

Horstmann (1973 a), in a meticulous study of within-caste variation of *Formica polyctena* found three behavioral categories within the worker caste. (1) Workers that climb onto trees gather honeydew, either from aphids or from other workers. Moreover, these workers also bring home about half of all insects. In spring, when insect harvest is at its peak, specialized insect hunters recruit from these tree visitors. (2) There are permanent specialized insect hunters all the time which forage on the ground and provide the colony with the other half of all insect prey. (3) Other workers are engaged in carrying home nest material which they gather exclusively on the ground. Beside these categories of workers, so-called "Leergänger" (non-carriers) seem to exist which are not engaged in any apparent labor and whose function is still conjectural. Specialization varies as a function of both collecting site and of labor performed: Honey-dew carriers are the most liable to change labor and insect-hunters foraging on the ground exhibit a more stable propensity to collect insects than those from trees. Workers revisit the same collecting site time and again, with 80 – 90% of them still visiting e.g. the same individual tree after four weeks. Polyethism, however, is specific to the labor per-formed and not a consequence of site-fidelity; otherwise both nest material gatherers and insect hunters could not jointly utilize the ground around the nest. Furthermore, honey-dew gatherers returning from their trees hardly make use of the opportunity to collect nest material or prey insects despite ample opportunity to do so.

In another study Horstmann (1973 b) could relate these different labors to body size as measured by head width (Table 13), following previous studies (review Wilson, 1972, p. 161 seq.). As can be seen, size of prey or nest mate-rial carried increases with body size although individual variation is large. Furthermore, workers harvesting honey dew are smaller sized than those collecting insects ($p < .03$). Apart from this correlation workers that roam far from the nest are larger than those staying close by, regardless of whether they search on trees or over the ground. Two equally speculative explana-tions have been proposed to account for the correlation between body size

Table 13. Mean head width and its frequency distribution in *Formica polyctena* workers in relation to the length of prey or nest material carried to their nest. Differences between random sample and all types of carriers ($p < .0001$) and between carriers of smallest and those of largest prey ($p < .0005$) highly significant. (From Horstmann, 1973 b)

Ants entering their nest	n	mean	frequency distribution(%)				
			1.1	1.3	1.5	1.7	1.9
Random sample	120	1.43	17	27	38	17	1
With prey $<$ 1 mm long	33	1.60	0	3	49	39	9
With prey $<$ 1 cm long	98	1.63	1	8	29	51	11
With prey $>$ 1 cm long	54	1.70	0	0	18	65	17
With nest material $<$ 1 cm long	31	1.56	0	10	55	35	0
With nest material $>$ 1 cm long	23	1.61	0	13	30	44	13

and radius of activity. One invokes a superior ability to orient towards the nest in larger workers which, at the same time, have larger brains. The other hypothesis simply suggests that stamina to cover long distances increases with body size (refs. in Horstmann, 1973 b). As yet, no decision as to the relative importance of both factors is possible.

Several species of Darwin's finches (*Geospiza* sp.) vary markedly with regard to beak size. As discovered by Bowman (1961) the large-beaked individuals husk, on average, larger seeds than smaller-beaked individuals. It is reasonable to assume that individuals have learned which type and size of seed to husk most efficiently. Such an explanation, however, can be confidently ruled out when considering division of labor and body size of *Formica* workers reported above: Workers with equally large head widths harvest nest material or large insects, and those with equally narrow head widths gather honey dew or small insects or short pieces of nest material. Thus there must be more specific determinants of behavioral specialization which cannot result from individual learning in relation to physical efficiency though other types of learning may well apply.

Morphological differentiation within a species in relation to differential foraging may be obligatory as evidenced by the respective sexual dimorphism in certain birds (Selander, 1966). Or, individuals facultatively transform structurally so as to cope with a new type of food if the need arises (3. B.). A most noteworthy transformation occurs in the tadpoles of spadefoot toads (*Scaphiopus*). Owing to proximate factors that are unknown, a minority of the tadpoles turn from alga-eaters into cannibals. Spadefoot tadpoles develop in ephemeral pools that dry out quickly and, as an apparent adaptation to this unstable environment, undergo the quickest metamorphosis known to occur among anurans (Bragg, 1966). Shortly before leaving the pool some of many *S. bombifrons* tadpoles may separate from the others and begin to prey upon their fellows that now form dense, well-oriented schools. Predation is facilitated by radical, irreversible changes of the mouthparts which develop a beak in the upper jaw which fits into a corresponding notch in the lower so that formidable teeth result. With these conspecifics are seized, torn to pieces and devoured (Fig. 66). In association with the beak/notch relationship the jaw muscles hypertrophy, and there are still other morphological changes. Variation from giant cannibals on one hand, intermediate forms, through the normal-sized noncannibalistic tadpoles on the other hand seems continuous. There is an indication that the tendency to form cannibalistic individuals

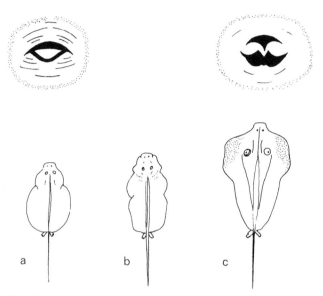

Fig. 66 a – c. Variation of body size, jaw size and mouthparts in tadpoles of *Scaphiopus bombifrons* in relation to herbivory vs cannibalism. (a) Normal-sized, non-cannibalistic alga-eater; (b) Intermediate with slightly modified mouthparts (not depicted), partially cannibalistic and preying upon *S. couchi* tadpoles; (c) Predaceously cannibalistic and preying upon all other kinds of tadpoles, with most extensive modification of mouthparts: Jaw muscles stippled; *Upper row:* Papillary fringes stippled; body length 61.5 mm; tooth rows represented by *lines*. (Redrawn from Bragg and Bragg, 1959)

varies between consecutive spawnings and perhaps between different progeny. Moreover, there is reason to assume that the behavioral and structural modifications of the cannibals are due to different genes as (1) only part of the intermediates show cannibalistic tendencies (Bragg and Bragg, 1959); (2) a substantial portion of the normal-sized noncannibalistic tadpoles kill and eat *S. couchi* tadpoles and thus behave in the same way as part of the intermediates although definitely not cannibalistically (Bragg, 1964). The existence of continuous variation suggests that mutual exclusiveness of foraging behavior and morphological outfit only holds for the extremes of the variability spectrum; the intermediates would appear to qualify as generalists as they feed on algae and conspecifics. Hence the picture emerging from this variation duplicates the one obtained for the oystercatcher, i.e. a combination of models *A* and *B*. The species forms a mixtum compositum of non-over-lapping specialists and of a large array of intermediate generalists whose feeding characteristics encompass those of both specialists.

Cannibalism appears to speed up metamorphosis as the consumption of any dead tadpoles already markedly speeds up growth (Bragg, 1960, 1962, 1964, 1966). This may be of importance in the hot parts of the range where pools dry out with dangerous rapidity. Facultative predation of *S. bombifrons* therefore seems to represent but one of several adaptive strategies of amphibians inhabiting uncertain environments (see above and Wilbur and Collins, 1973).

5. *Predatory Versatility in Relation to Prey Availability*

The foregoing account has demonstrated that predator individuals when foraging may differ in tactics and/or prey species consumed. It seems further that extensive overlap (model *B*) between individuals is the rule and preda-tory discreteness (model *A*) the exception. Moreover, even where specialists occur, other individuals are generalists; these "jacks-of-all-trades" are versa-tile enough to cover all the single skills of their specialist conspecifics, as exemplified by anoles, oystercatchers, ant workers, and spadefoot toads.

The question arises whether predatory overlap is more common because it is more successful in evolutionary terms. Both models of predatory versatility assume that each individual uses some tactics more or consumes some prey animals more than others. It is reasonable to assume that such preferences reflect hunting efficiency. If prey is abundant the entire range of versatility can be exploited by specialists and generalists alike. On the basis of partially extensive field work it has been suggested repeatedly that during food shortage the spectrum of feeding patterns contracts so that each individual exercises only those with a maximal return of energy, i.e. those which it employed most frequently anyhow; it would have to give up the less efficient part of its repertory. As feeding conditions improve diversity of exploitation patterns again expands so that pulsations in accordance with changes in food supply result (Lack, 1946; Gibb, 1954; Nilsson, 1960; Lindstrom and Nilsson, 1962; Root, 1967; Zaret and Rand, 1971). As pointed out by Klopfer (1968, p. 25) food shortage including interspecific competition for food, would affect a species of specialists (model *A*) differently than a species of generalists (model *B*). Whereas a generalist is able to contract diversity, specialists can hardly follow suit because their range of tactics is already rather limited. And if food shortage or competition is extensive enough to affect all the food items which one or even many specialists normally exploit contraction is of no avail to them and they would die. If this is true, discreteness of intra-specific predatory specialization must confer advantages that outweigh the lack of individual adaptability. The argument, however, needs some qualification: (1) Food shortage need not be a universal threat. It is hard to believe that non-cannibals and cannibals among spadefoot tadpoles feeding on algae or conspecifics, respectively, would ever run short of food. Furthermore, oystercatcher stabbers and oystercatcher hammerers are not known to differ in their diet, so that shortage of any prey animal would affect both of them equally. By comparison, specialization concerning predatory patterns which are tuned to certain prey animals (Chap. 5. B. I.) is a more serious risk in terms of food shortage. (2) Specialist anoles and snakes have been found to change their original prey preferences either for reasons still unknown or as a consequence of being fed a non-preferred prey. The adaptive significance of this developmental change is still obscure. One may speculate that specialization is effective enough to achieve optimal hunting success and to mitigate intraspecific competition under certain conditions; at the same time specialization is not too rigid to prevent the predator tiding over the shortage of a particular and originally preferred prey species.

The brief account of the existence of pulsations in time of habitual predatory specialization is at variance with a considerable body of facts and well-reasoned theory. As will be recalled the Syrian woodpecker has been found to use diverse food searching movements more evenly when hungry than when sated (Chap. 1. A. V.). Likewise, two bird species have been found to expand predatory diversity when feeding their brood as compared to self-feeding. This is achieved either by employing more diverse hunting maneuvers and/or by extending foraging trips into sites normally not exploited (Chap. 1. B. III.). Both findings can be accomodated by the assumption that with growing food demands selectivity of prey selection drops so that visitation of e.g. more feeding sites yields more food per unit of effort. This conclusion is in accord with the other finding that with growing hunger, predators become less and less selective to the range of prey stimuli and thus enlarge their food spectrum (Chap. 1. A. III. 1). Accordingly, there is a concomitant increase of predatory diversity (termed above "versatility"). Similarly, by his "optimal foraging model" MacArthur (1972, p. 59 seq.) was able to predict that a species should become more specialized ("less versatile") in a productive than in a harsh environment. The rationale behind this prediction rests on a consideration of a predator trying to maximize its food intake per unit time. On locating a prey item, a predator has to decide whether to pursue or to search again for a better item and pursue it instead. Since either choice ends with the animal ready to start a new search, the better decision is that which promises a higher yield per unit time. Hence, an animal should elect to pursue an item if, and only if, during the time the pursuit would take, it could not expect both to locate and catch a better item. In other words, a new item should be caught only if the pursuit time needed is less than the average search plus pursuit time for all previous items. In a harsh environment with a low prey density search time is large. Therefore pursuit time for nearly all items encountered is immaterial in terms of decision making and the animal has to become a generalist. Conversely, in a productive environment with a small search time specialization is favored. (Interestingly, these reflections yield another prediction: a searching predator should be a generalist since its search time generally by far exceeds its pursuit time. Conversely, a "pursuer" – with prey always in sight, so that search time is negligible, but not pursuit time, should be a specialist.) To support his model MacArthur (1972, p. 62) points to observations on great blue herons (*Ardea herodias*). In productive Florida waters these selected a much narrower range of food size than they ate in the unproductive lakes of the Adirondacks.

From this sketchy account of fact and theory it becomes clear that the above-mentioned view of the pulsation of facultative predatory specialization and its functional interpretation needs some definite qualification. It seems that the conditions conducive to this specialization need to be defined carefully in each single case. One possible way to resolve the discrepancy would lie in a specification of the meaning of "expansion of the feeding niche". If it merely involves an incorporation of new items to the diet, not necessitating new motor skills of capture or disposal niche expansion may be relatively easy. If, however, it involves new types of performance of food acquisition physiological constraints to achieve this may well prove prohibitive. The oystercatcher needs nearly a year to attain peak performance with a single skill. Again, different sorts of performance almost certainly pose different difficulties: Adding a new searching site to the old ones is presumably easier to achieve than to acquire a new capture technique; in copying both from conspecifics and alien flock members tits have been shown to need just *one* experience to broaden their spectrum of searching sites and hence predatory versatility (Krebs 1973 b, Krebs, MacRoberts, and Cullen, 1972; see also Chap. 2. C. IV. 2.).

C. Behavioral Aspects of Hunting Success

Hunting success as measured by the ratio of actual captures per hunting attempt is a measure most important in one of several respects. It permits judging most directly on the adaptedness of hunting methods (as well as of escape tactics of prey).

I. A Comparison of Hunting Success across Predator Species

Predators that habitually capture evasive prey are usually less than 80 – 90% successful (Table 14). The synopsis of hunting success rates needs, however, some qualification. Stalking predators usually embark upon many more hunts than eventually develop into actual rushes or chases. This pertains especially to the large carnivores with their keen sense for the vulnerability of prey. As stressed by Schaller (1972, p. 241), a chase may develop from the faintest glance at, or an unobtrusive trot towards prey, an abortive stalk etc., with many subtleties of the situation determining the course of events. The data included in Table 14 reflect success rate as measured for "determined" or fast chases and could have yielded considerably lower estimates if all attempts had been tallied; this may be particularly true for the avian and mammalian predators included. However, the exclusion of all aborted hunts lends some confidence to an interspecies comparison.

Table 14. Hunting success of various predators. For further carnivores see Schaller (1972, pp. 446, 451, 455)

No.	Predator	Prey	No. attempts	% successful	Comments	Reference
1	*Woodruffia metabolica* (Holotricha)	*Paramecium* sp.	many	14	2.2%/encounter; attempt = dilate mouth to engulf a *Paramecium*	Salt (1967)
2	*Busycon carica* (Gastropoda)	*Venus mercenaria*	26	58		Carriker (1951)
3	Cuttlefish	Shrimp	?	ca. 90		Messenger (1968)
4	Largemouth bass	Fish	85	94	Prey fishes without evasive movements	Nyberg (1971)
5	Forster's tern (*Sterna forsteri*)	Fish, ca. 4 spec.	1538	24		Salt and Willard (1971)
6	American kestrel	Rodents (insects?)	?	33	On familiar hunting ground	Sparrowe (1972)
7	Osprey	Fish	469	80–96	Both dives and snatches from surface	Lambert (1943)
8	Various raptors (*Falco columbarius, F. peregrinus, Accipiter nisus, Haliaetus albicilla*)	Birds	688 (60–260)	7.6 (4.5–10.8)	On migration	Rudebeck (1950, 1951)
9	Black bear	Salmon	1481	38.6	During 310 fishing sequences	Frame (1974)
10	Wolf	Moose	77	7.8	From all moose 'tested' (p. 129)	Mech (1970, p. 246)
11	Spotted hyena	Wildebeest calf	108	32	Similar for Wildebeest adult, Thomson's gazelle, zebra	Kruuk (1972 b, p. 172)
12	Puma	Deer, elk	45	82	Excluding aborted hunts	Hornocker (1970)
13	Cheetah	Thomson's gazelle	87	70	Only fast chases tallied	Schaller (1972, p. 455)
14	Chimpanzee	Mammals, 6 spec. Olive baboon (*Papio anubis*)	95 18	40 36	Including primates Adolescent victims < 2 yrs	Teleki (1973, p. 57) Teleki (1973, p. 116)

194

The initiation of the final attack phase requires both precise information on the location of the prey and on the predator's own velocity. In this regard it is noteworthy that the failure rate hardly falls below 10% in predators as diverse as cuttlefish, bass, osprey, and perhaps puma (Table 14, Nos. 3, 4, 7, 12). As pointed out by Nyberg (1971), prey fishes when attacked by his largemouth bass had often not shown any evasive movements. In those cases therefore any miss had necessarily resulted from inadequacies inherent in the sensory-motor machinery underlying attack (see also Nursall, 1973). Remembering that the last attack phase is internally preprogrammed in that it is based upon sensory information prior to its onset, this information must produce motor output in tens of msec (p. 145). The errors of the bass' preprogramming system with stationary prey were largely those of timing, e.g. the bass opened its mouth too soon. They were defined by Nyberg (1971) as causing the "intrinsic failure rate" which in the bass is approximately 6%.

It is then logical to define as the "extrinsic failure rate" the portion of all misses resulting from external interference with a predator's attack, notably actions by the prey, regardless of whether they were induced by the predator's approach or not. The total failure rate, the parameter observable in general, results from the combined effect of both the "intrinsic failure rate" and the "extrinsic failure rate". Hence, given an estimate of the "intrinsic failure rate", one could obtain from observations of the total failure rate an estimate of the efficiency of the prey's defence system. Its efficiency would be proportionate to the "external failure rate" if other sources of external interference with attack can be excluded.

It is noteworthy that the low success rate of four diurnal raptors (Table 14, No. 8) of 7.6% contrasts with the usually higher success of other avian predators, even including another hawk, the American kestrel (Table 14, Nos. 5, 6, 7). A possible explanation for this disparity is that all four former species were on migration whilst the latter were roaming over their familiar hunting grounds though in the case of the fish-eaters this will not mean very much. A more important reason might be that prey birds are much more evasive than prey fishes and rodents.

Chimpanzees have been discovered to hunt regularly and successfully after various mammals of their habitat, though meat constitutes less than 1% of their total diet (Teleki, 1973, p. 58). Chimpanzees therefore qualify as "omnivorous predator-foragers" and the motivation for chasing and killing other animals is far from being clear (see p. 201). It is noteworthy that all olive baboons that were captured by chimpanzees were subadult animals (Table 14, No. 14). Adult baboons are too wary and too dangerous to be tackled, a point which will be discussed below.

Despite often very low hunting success predators reserve surprisingly much time for resting. *Amoeba proteus, Woodruffia metabolica* (Salt, 1967), African fish eagles (Brown, pers. comm.), lions, and wild dogs (Schaller and Lowther, 1969) spend about ⅙ of their time in hunting and eating, and oystercatchers even only ¹⁄₁₂ (Drinnan, 1957 quoted by Salt, 1967). An ambush predator, a mantid, took only some 4% which was spent almost wholly in eating (Holling, 1966).

II. Variables Influencing Hunting Success within Predator Species

For the sake of simplicity hunting success has been dealt with thus far as a feature characteristic for each predator. In addition, hunting success varies greatly with the type or "setting" of the prey, especially its locomotory or escape abilities (Table 15). For instance, flying flies are more difficult to catch

Table 15. Hunting success in relation to type of prey

No.	Predator	Prey	No. of attempts	% successful	Comments	Reference
1	*Hierodula crassa* (Mantidae)	Fly walking	898	63.0	70.2% strikes/contact	Holling (1966)
		Fly flying	112	13.4	33.1% strikes/contact	
2	Herring larva, 35–42 days	*Artemia* nauplii	81	100		Rosenthal (1969 b)
		Larger plankton, *A.* metanauplii	303	96.5		
3	Red fox	Rodents	ca. 58 [a]	25 –	Depending on snow conditions. [a] 162 km tracked; [b] Roe deer (*Capreolus capreolus*), European hare (*Lepus europaeus*), red squirrel, birds	Palm (1970
		Larger prey [b]	9	100 0		
4	Wild dog	Thomson's gazelle < 2 mo	22	95	Mean = 70%, incl. other prey except zebra	Schaller (1972, p. 451; see also Estes and Goddard, 1967; H. and J. van Lawick-Goodall, 1970, p. 61)
		Thomson's gazelle > 2 mo	47	49		
5	Cheetah	Thomson's gazelle fawns	31	100		Schaller (1970), Schaller and Lowther (1969)
		Thomson's gazelle adults	56	54		

[a, b] See under comments.

196

for a mantid than are walking flies (Table 15, No. 1). The probable reason for this is the higher overall velocity (see also Table 15, Nos. 2, 4, 5) in flight and the absence of pauses which are so characteristic of a fly when walking. Further, although success rate in herring larvae varies little with prey type (Table 15, No. 2) the difference shown seems to be significant; the lower value applies to larvae on average four days older than the perfectly successful fellow members, although capture success increases markedly with age, i.e. from about 10% at 5 days to about 96% at 41 days of age, with little improvement later on (Rosenthal, 1969 b).

In addition to the carnivores listed (Table 15, Nos. 4, 5), spotted hyenas also probably have much less difficulty in catching Thomson's gazelle fawns than older animals (Kruuk, 1972 b, p. 194). Not only are fawns easily overtaken while running, but they also are often picked up by searching predators, including olive baboons (Harding, 1972), without having any opportunity to flee. When left alone by their mothers, fawns crouch in tall grass and rely entirely on their camouflage.

Apart from the particular prey, environmental conditions may also influence hunting success as suggested by a study of Palm (Table 15, No. 3). Deep snow may markedly reduce the rate at which red fox can catch mice. The striking variation of capture rate from 25 – 100% on different nights is not readily explained. Perhaps snow conditions at times favored the mice and at other times the fox. Although the sample is small the study also suggests that prey with well-developed long-distance sense organs and high powers of escape may doom hunting efforts futile. The fact that the fox attempts to capture such evasive prey at all suggests once more that a predator takes a chance even if the prospects of a hunt are poor (see p. 179).

Apart from external factors discussed so far the condition of the predator may be important too. Injuries through intraspecific fighting or hunting impair hunting success of lions (e.g. Schaller, 1972, p. 44 seq., p. 189 seq.). The tiger, which generally spurns man, may then take to man-eating though other causes contribute to this habit (Schaller, 1967, p. 278). Furthermore, predators become enfeebled by old age. For example, baggy, fat old hyenas can no longer run swiftly after prey (Kruuk, 1972 b).

An important point borne out by Table 15 is that 100% capture rates *do* occur. This means that the "intrinsic failure rate" may become zero in certain cases. It should be stressed however, that this is clearly a function of the evasiveness of the prey sought: *Artemia* nauplii move erratically and do not gauge their movement to the proximity of a herring larva (Table 15, No. 2); and gazelle fawns do not offer much difficulty to cheetah which overtakes them within seconds (Table 15, No. 5). Similarly, Rudebeck (1950, 1951) attributed to kestrels (*Falco naumanni, F. vespertinus*) a 100% capture success with swarming termites. This contrasts with the very modest capture rate of raptors with much more elusive prey (Table 14), including desert locusts of which kestrels (*Falco tinnunculus, F. naumanni*) caught only 74% when stooping (Stower and Greathead, 1969). From this enumeration it would seem to follow that the intrinsic failure rate drops to zero if the prey's velocity and/or its sensorial powers are substantially inferior to those of the predator. In the light of this suggestion the occurrence of misses by largemouth bass when attacking stationary fish (Table 14) appears puzzling. These misses are presumably best explained by assuming errors in the internal control of attacks. Hunters which, like mammalian carnivores, at times attain a 100% success rate also miss prey initially but thereafter can correct for an initial miss which, hence, will not show up from the data; these merely mirror the final outcome of a hunt. Thus an intrinsic failure rate above zero would possibly appear more widespread if initial attacks on easy prey could be documented more often.

As already stated there are many more incipient hunting attempts than "true hunts" and the discussion so far was concerned only with these. It is noteworthy that hunting success increases in some way with the tendency to attack as measured by the ratio of true attacks/contacts with prey. Two points must be made clear.

First, the tendency to attack may increase with the vulnerability of the prey, or with the "prospects" for success, as is borne out by mantids which not only succeeded in getting more walking than flying flies but which also struck at walking flies more often than at flying ones (Table 15, No. 1). Here then we arrive at a quantitative measure of how the predator judges the "prospects" of its attack; given the same accuracy of strike in both situations, in terms of the predator's effort, one can surmise that the mantid assessed the prospects of striking at a walking fly as superior to striking at a flying one. The tentative conclusion reached thus far (p. 179) needs now to be refined: whilst it is true that predators take a chance for attack, even if the prospects for success are poor, they do so apparently less often than if their chances are better.

Second, the tendency to attack may increase with age as, for example, in the herring larva. At the same time capture success rate increases too, but much more rapidly than the tendency to attack. Moreover, attack tendency increases irregularly while success rate does so almost constantly (Rosenthal, 1969 b). This developmental change then seems to suggest that attack tendency and hunting success are governed by different developmental processes. Furthermore, hunting success may feed back upon attack tendency. Another influence is hunger. As reported above for a mantid (p. 20), hunger enhances the tendency to attack but does not affect strike accuracy. Therefore strike accuracy must be buffered against changes of attack tendency inasmuch as these are due to changes in hunger.

Correlations between the defense of the prey or the hunting situation and particular hunting techniques hinted to the adaptedness of the latter (Chap. 5. B. I., II.). Yet there is a dearth of information as to how hunting success relates quantitatively to hunting method. Schaller (1972, pp. 254, 445) noted lions employing eight different hunting techniques almost each of which produced a very different result. For example, single lions are twice as productive when stalking prey than when running (18 vs 7%). If one assumes that lions can assess this difference it becomes puzzling why they do not always choose the most successful technique. It may be relevant in this context that the proportions of hunting techniques vary only slightly with type of prey. A likely explanation, therefore, would be that lions compromise between the circumstances of a hunt, e.g. proximity of cover, and the most productive technique; under unfavorable conditions, e.g. in absence of cover, they choose an inferior technique like running which, however, is presumably the best available. Similar considerations apply to the difference in hunting success between single and group hunting (see below).

Spotted hyenas use diverse hunting methods geared to different prey animals. For example, they hunt zebra in groups of over 20 animals but gazelle singly. Despite the diversity of methods hunting success is surprisingly similar: hunting wildebeest adults was successful in 44% of the observations, calves in 32%, zebra in 34%, and gazelle in 33%. Accordingly Kruuk (1972 b, p. 205) suggests that hyenas are able to "gauge the chances of securing the quarry" by resorting to the most appropriate method each time. Thus both the lion and the hyena study suggest basically the same conclusion though for different reasons.

Earlier in this book it was concluded that the levels of motivation underlying hunting permit its occurrence much more often than would appear necessary in terms of the predator's food requirements (Chap. 1. B. I.). A major selective agent, which has moulded the causal mechanisms in this way, is now believed to lie in a success rate which is almost universally less than 100%: the readiness to hunt must be high enough to permit a large proportion of failures to occur.

III. Aspects of Communal Hunting

Communal hunting may crucially affect hunting success and the size of the quarry attacked. In the following both the reasons for these differences in hunting success and the benefits of group hunting accruing to the individual will be examined.

1. Modes and Properties of Communal Hunting

Degrees of Cooperation. Communal or group hunting denotes the concerted effort of at least two predators to track and hunt down prey. The fortuitous arrival of two conspecifics at a targeted prey animal at the same time, each of whom is able to dispatch it alone (e.g. Federighi, 1931; Galtsoff *et al.*, 1937) does not qualify as communal hunting as it is understood here. As a rule, group hunters belong to the same species. Whether the adoption of roles, if only temporary, during a hunt (5. B. III. 2.), is bound to species with a particular group structure remains still an open question.

In the simplest case communal hunting implies the simultaneous efforts of group members to secure prey, mostly from a number of elusive prey items available. Each member is supposed to benefit from attacks of its neighbors in that these cause prey to flee that is right within their reach, each individual unwittingly acting as the "beater" for others. Bartholomew (1942) found that large groups of 500 or more double-crested cormorants (*Phalacrocorax auritus*) took up a different formation and were noticeably more successful than small groups of fifty or less. For this facilitatory effect to occur, the efforts of group members need not be gauged to those of others (Sette, 1950: *Pomatomus saltatrix;* Nursall, 1973: *Perca flavescens;* see also Bullis, 1961). In cases such as these it is as little necessary to postulate any elaborate organization of the group as in mixed species flocks whose members mutually act as beaters (further examples see Chap. 5. A. I.).

Close proximity of conspecifics is by no means a guarantee for hunting success to improve. For unknown reasons protozoa (*Amoeba proteus, Wood-*

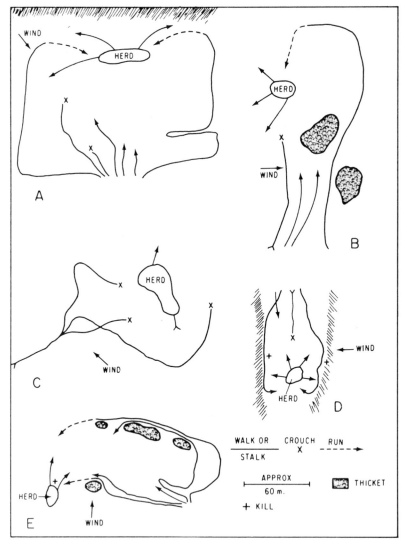

Fig. 67. Five examples of routes taken by lionesses when stalking prey cooperatively. (From Schaller, 1972, p. 252, by courtesy of the Univ. Chicago Press, © Univ. Chicago)

ruffia metabolica) capture less prey as their own species density increases (Salt, 1967). Purely mechanical reasons, such as collision with conspecifics in mid-air, adventitiously decrease hunting success in Forster's terns (Salt and Willard, 1971) and sparrowhawks (Meinertzhagen, 1959, p. 94). Moreover, social hostility interferes drastically with egg stealing of carrion crows in gull colonies; cooperating pairs of crows thus suffer already from the presence of one extra crow (Kruuk, 1964, p. 28; see also Salt and Willard, 1971).

Communal hunting may take the more elaborate form of cooperation. Here then group members tune their actions to what others do when approaching or when attacking prey. A number of predatory species "herd" their prey by approaching it with the pack fanned out widely; members at the ends of the front approach with greater speed than those in the middle so that the prey eventually gets encircled or is driven into a cul de sac with small prospects of escape (see p. 149). "Herding" has been seen to work in certain sharks (Bigelow and Schroeder, 1948 quoted by Breder, 1967; Eibl-Eibesfeldt, 1962), in *Caranx adscensionis* (Eibl-Eibesfeldt, 1962 b), killer whales when pursuing pinnipeds (Evans and Bastian, 1969; Martinez and Klinghammer, 1970), porpoises (*Phocaena vomerina*) (Fink, 1959 quoted by Hobson, 1968), and in wolves (Mech, 1970); perhaps in pelicans (*Pelecanus* sp.) too, when they fish buoyantly as a compact party and with bills depressed, dipping their heads under water in unison (Landsborough-Thomson, 1964, p. 607). Similarly, wattled starlings (*Creatophora carunculata*) form a huge flock in the air and, like an erected cylinder, encircle locust swarms to feed on them (Friedmann, 1967).

Another variant of herding is "driving" in which a few hunters walk around a herd of prey in order to drive it towards others waiting in ambush as practised by wolves (Murie, 1944; Crisler, 1956; Kelsall, 1968 quoted by Schaller and Lowther, 1969) and lion prides (Fig. 67); each animal is patterning its behavior after that of the others to the extent of obviously looking at neighbors during a stalk (Schaller, 1972, p. 250), or in the case of wild dogs, during full chase (Kühme, 1965). Wild chimpanzees, when intent on hunting a small mammal as a group, stare at it in a tense posture of alertness. In between they stare at each other. By these "predatory stares" they obviously synchronize pursuit, for its onset is explosive with all hunters trying to forestall escape of the quarry simultaneously (Teleki, 1973) [14].

In *A*, *B*, and *D* of Fig. 67 the encirclement of lions had succeeded and one or two lions caused prey to run toward the others which were waiting; in *C* the herd moved off before the flanking movement of the stalk was completed; and in *E* two lionesses rushed before three others were fully in position.

---

[14] Whether this behavior represents true predation is conjectural if the latter is thought to aim solely at procuring meat. Both Kortlandt (1972) and Teleki (1973) agree, though with different emphasis, that chimpanzee "predatory" behavior may serve several functions. Kortlandt (1972) suggests that it may function as a systematic control of food competitors (baboons pay the largest toll) and as an intra-group intimidation display. Any theory concerning the motivation and function(s) of chimpanzees' "predatory" behavior has to take into account its puzzling absence west of the 29th meridian which has not been satisfactorily explained either.

During such hunts lions integrated their actions solely by observing each other's posture and movements; neither sounds nor facial expressions were used which, at any rate, would not have been useful at night. If only one quarry is involved, all endeavor to catch it. But if there is a herd, cooperation ceases as soon as each tries to capture the nearest one (Schaller, 1972, p. 250 seq.). By being widely spread out, lions increase (1) their chances of coming into contact with prey which may either scatter and thus run at a hidden lioness (lionesses doing most of the hunting); (2) by fanning out, lions diminish the chance of giving an undesired warning to the herd aimed at. Cooperative night hunts proved almost twice as (42%) productive as day hunts (27%). Apparently darkness, though moon-lit, benefited the lions more than their ungulate prey. As asserted by Schaller (1972, p. 253) dark nights would have enlarged the difference in success still more, a remarkable possibility when considering that communication between pride members is entirely visual (see above).

It should be noted that encirclement need not involve, as with this carnivore, group members being aware of the consequences of their actions in relation both to others and to the prey. Awareness of the other's action alone already offers an advantage as it diminishes the prospects of successful escape of prey. This seems to be the case with pelicans when cooperatively encircling fish which they either do not or only partially see. Running predators such as wild dog have been said to employ "relay hunts" in which exhausted pack members are replaced by others which have trailed behind (Estes and Goddard, 1967; Schaller, 1972, p. 340). This is not reported, however, for other coursers like hyenas or wolves (Kruuk, 1972 b; Mech, 1970). It is not known whether, as the individual leading the chase falls back, the others continue at their original velocity, or speed up as a response to the leader falling back. Either way would result in a "relay hunt".

Another important feature of pursuit of fleeing prey is that pack members lagging behind often succeed in seizure if the targeted prey veers off to one side and they can cut corners, a technique which is of less avail to the chase leader (Estes and Goddard, 1967; Schaller, 1972; Schaller and Lowther, 1969; Mech and Frenzel, 1971; Hatch, manuscript: gulls). In much the same way as many prey individuals can confuse predators by protean displays, a group of predators can obviate coordinated defense of prey animals by "multiple lure displays" which are equally protean in nature (Humphries and Driver, 1970). The erratic succession of the attacking individuals is the protean aspect of the display. The prey's attention is thus repeatedly distracted so that it can seldom launch a counter-attack (for example see Chap. 5. A. V.).

Apparent "multiple lure" cooperative hunts have also been observed in pairs of crows by Axell (1956), when about to steal an egg from a herring gull's nest, and in many birds of prey (Berndt, 1970; Bräuning and Lichtner, 1970; Brown, 1970); groups of three also occur (Gordon, 1955, p. 94). Whether the function is always, as is likely in the latter case, confusion of the prey, or strengthening of the bond, as suggested by Brown (1970) for certain eagles, remains to be studied. If the size of the quarry permits, African fish eagle mates share it with one another (Brown, pers. comm. 1972). In which way cooperation affects hunting success will be reported upon below.

The Adoption of Roles. Members of a group partaking in a cooperative hunt are known to adopt different roles when tackling a quarry. For example, a pack of dwarf mongooses when attacking a snake acts as a multiple lure display. A "strike instigator", and later others, lunge toward the snake's head while the "disorienter" grabs it by the tail end and momentarily releases the victim so as to offer no target for a retaliating bite. This "disorienter" also later on kills the snake (Rasa, 1973). It is generally unknown how persistent the adoption of such roles is (see 5. B. III. 3.). But mates of Mediterranean lanner falcons consistently adopt very different roles when hunting together. The heavier female chases birds [jackdaws (*Coleus monedula*), rock doves (*Columba livia*)] from a seaward cliff by flying in and out of caves and crannies; meanwhile the male is waiting on the wing some distance from the cliff. In this situation the female persistently acts as the "beater", its mate as the "hunter" (Glutz et al., 1971; p. 833).

A much more complex state of affairs holds for the predatory behavior of wild chimpanzees according to the well-documented work of Teleki (1973). Predation upon small mammals is commonly, though not entirely, a social affair, which is moreover cooperative (see above). Almost only males do the hunting but the carcass is distributed amongst the whole band including females and children. Whereas in everyday life an alpha male dominates all other males of the group any male can take over a "control role" during pursuit of a targeted prey animal. The male that is closest to the targeted or pursued quarry initiates or continues the hunt, so that several group members in succession may adopt the "control role". The male which during possession of the control role, finally seizes and kills the victim is even allowed to begin dismantling and distributing it. It is important to note that such a flexible role system facilitates capture, as it forestalls interference from the alpha male. Similarly, wild dogs (Kühme, 1965; Estes and Goddard, 1967) and wolves (Mech, 1970), while chasing a quarry and communally feeding upon it are completely peaceful towards each other, although a definite rank order exists at other times (Mech, 1970; Fox, 1972); in wild dog packs, however, it is recognizable least among all socially living carnivores (Kruuk, 1972 b, p. 279; but see H. and J. van Lawick-Goodall, 1970, p. 95).

2. Factors Conducive to Communal Hunting

Communal hunting is obligatory, as in the wild dog (for few exceptions see Kruuk, 1972 b, p. 279), while it occurs only occasionally in other predators as, for instance, gulls (p. 208) and spotted hyenas. In these latter carnivores, however, the sociability needed for cooperation to become effective may be flexibly geared to varying food supplies. Whereas nomadic singletons and small groups inhabit the Serengeti area, large clans form the basis for the hunting parties of varying sizes in the Ngorongoro Crater. The different social organization was interpreted by Kruuk (1972 b, p. 268) as long-term adaptive responses to average prey size, which in the Serengeti is usually small, e.g. Thomson's gazelle, and in Ngorongoro large, e.g. many zebra and wildebeest. A similar, though more complex consideration applies to seasonal and longer-term changes of social organization of Serengeti lions (Schaller, 1972, p. 357).

Apart from the more immediate aggregative responses of hyenas to the external stimuli of the prey animals around them, internal factors seem to play a role too. As described earlier (p. 31), members of a clan indicate by their behavior prior to a hunt what species they intend to kill. When going on a zebra hunt often more than 20 animals assemble at the den site whereas only a small fraction of this number sets out when hunting wildebeest or gazelle. Moreover, the pack may pass many otherwise suitable prey on their search for the intended prey, thereby indicating that they anticipate finding a particular prey. Besides, there are marked differences between clans with regard to the preferred prey. It was argued in some detail that internal factors rather than prey availability determine pack size, search behavior, and prey preferences.

3. Benefits of Communal Hunting

Numerical Hunting Success. Most recently communal hunting was recognized to play a key role in hunting success. For example, groups of lions kill about twice as many prey per hunt than single lions (Table 16). The data for the prey species listed are similar except that there is an unexplained drop with three lions hunting zebra. Large lion groups do not catch prey more successfully than small ones but there is an increase in the actual number of animals captured, something not showing up in Table 16. Wildebeest are especially subject to multiple killing by groups of four or more lions (Schaller, 1972, p. 254). A similar increase of hunting success with increasing group size occurs with gulls (p. 208), jackals, and spotted hyenas. The reasons for this

Table 16. The relation of hunting success to the number of lions stalking or running. (Modified from Schaller, 1972, p. 445, by courtesy of Univ. Chicago Press)

No. of animals hunting	Thomson's gazelle		Wildebeest and zebra		Other prey		Total	
	No. hunts	% success	No. hunts	% success	No. hunts	% success	No. hunts	% success
1	185	15	33	15	31	19	249	*15*
2	78	31	17	35	11	9	106	*29*
3	42	33	16	12.5	5	20	20	*27*
4–5	42	31	16	37	4	25	25	*32*
6+	15	4.1	21	43	7	0	0	*33*
Total	362		103		58		523	

important increase have been examined in detail for hyenas hunting wilde-beest calves (Kruuk, 1972 b, p. 174 seq.). As with lions, the main improvement occurs if two hyenas instead of a single one are hunting. Most of the success that results from two hyenas cooperating is due to their being able to avoid the attacks of the mother wildebeest once the calf has been overtaken in the chase. The wildebeest cow can take on only one hyena at a time, which enables the other hyena to deal with the calf. In Fig. 68 it is shown that if the

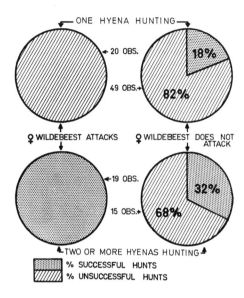

Fig. 68. The effect of attacks by wildebeest cows on hunting success of different numbers of hyenas chasing their calves. (From Kruuk, 1972 b, p. 175)

female wildebeest does not attack, two or more hyenas are more effective than one at catching a calf, but not significantly so; in this category calves avoid capture largely through outrunning hyenas or disappearing in larger herds. However, if the mother wildebeest does attack, a single hyena is 100% ineffective in catching a calf, whereas a pack together is 100% successful in doing so. These figures demonstrate the benefit that carnivores may derive from cooperation and, when taken together with the cow's behavior, show for the first time quantitatively the benefit of a multiple lure display. The data also indicate the inability of a female antelope to defend her off-spring completely against more than one predator at a time and show the survival value of active defence against singletons. An optimal strategy of the female wildebeest would consist of always attacking when facing a single hyena and of always running, when facing more than one, a goal which for wilde-beest is apparently difficult to attain.

Size of Prey. Apart from greater hunting success, communal hunting can confer the advantage of subduing a larger quarry than otherwise. This is already seen in some communally living spiders. *Agelena consociata* (Theri-diidae) lives in large webs up to 3 m in widest diameter along borders of the Gabon forest. "Numerous sheets are spun in an irregular manner by the hundreds of spiders that inhabit each nest. Small prey are captured and eaten singly by the individuals that happen to be closest at hand, but larger prey are pursued and killed by groups." (Wilson, 1972, p. 133 and Refs.). The spiders benefit by working in groups which are able to capture prey that would be too large for an individual to handle (Fig. 69). A second probable

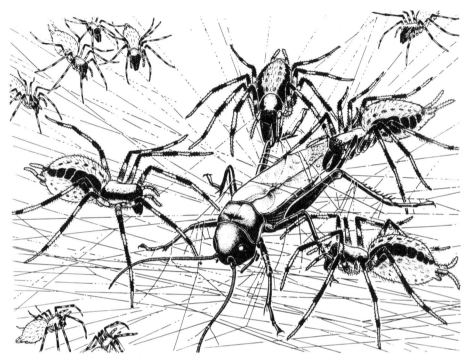

Fig. 69. Adults of *Agelena consociata* attack a cricket on their communal web, while the spiderlings wait nearby. (From Wilson, 1972, after photographs by B. Krafft, The Insect Societies, by courtesy of Harvard Univ. Press)

benefit is the added protection given colony members who can remain hidden deep within the silken retreats of the giant nests. A parallel to this group hunting is provided by the group predatory habits of ants. While many non-dorylines engage in group killing and retrieving of larger prey, many dory-lines group-raid on top of that, i.e. forage in well-coordinated assemblages. Group-raiding ants not only flush more prey than foragers acting independently but also can overpower much larger prey; a correlation that holds also for species of one and the same subfamily (Ponerinae). Wilson (1972, p. 68) suggested that the inclusion of larger prey into the species diet has been and is one of the key factors in the evolution of large colonies and of nomadism, so famous a feature of tropical army ants (Ponerinae, Dorylinae). According to experiments by Büttner (1973), the number of ants fighting a quarry rises linearly with its volume. In a comparable way, stamina of fighting increases

206

in relation to the strength of the quarry's struggling, whereas the number of squirts from the poison glands does not.

The killing of large prey animals by groups of species members is also a conspicuous feature of mammalian carnivores. As a rule, predators of the species involved can hunt singly and with success, but if alone, they do not try to attack the large prey typically pursued by the pack. Thus spotted hyenas go in groups averaging 1.2 when hunting Thomson's gazelle, 1.6 when chasing wildebeest calves, and 10.8 when starting a zebra hunt (Kruuk, 1972 b, p. 203). A similar state of affair holds for other social carnivores and their larger prey animals, e.g. for wolves and moose (Murie, 1944; Mech, 1970), cheetah with kongoni (*Alcelaphus buselaphus cooki*) and zebra (Schaller, 1970), lions and buffalo (Schaller, 1972, p. 260), jackals and adult gazelle (H. and J. van Lawick-Goodall, 1970, p. 138), and also Eskimo when hunting bearded seal (*Erignathus barbatus*) as opposed to ringed seal (*Pusa hispida*) and similar small species (Rasmussen, 1931 quoted by Schaller and Lowther, 1969). Killer whales, which habitually move in groups, focus their joint efforts when encountering single large baleen whale (Mysticeti) instead of smaller prey which pack members can tackle on their own. Their hunting behavior changes dramatically in immediate response to the former prey (Martinez and Klinghammer, 1970).

From the foregoing account it is clear that predators, by hunting communally, can subdue prey that would be invulnerable to every single one. Sharing a larger carcass among pack members raises the question of the nutritional benefit accruing to each member. Benefit would seem to depend on the size of the carcass in relation to individual sizes of a filling meal. For example, spotted hyenas increase in number with the weight of the carcass available for consumption. This increase may take place prior to a hunt, as with zebra, or during the chase, as with wildebeest. It is important to note that meal size (weight) per hyena also increases with the weight of the carcass. This means that hyenas do not gather in proportion to carcass size but less. Although there is no precise estimate of how much a hyena must eat in the long run to remain healthy, a conservative estimate, including data from different seasons of the year, is 2 kg per hyena per day (Kruuk, 1972 b, p. 76 seq.). If one compares this with the relation between weight of meal per hyena per carcass and the weight of the latter it becomes evident that already small prey animals weighing less than about 30 kg satisfy the minimum food requirements (Kruuk, 1972 b, Figs. 32, 33). One has to bear in mind, however, that this is due to the hyenas gearing their numbers to carcass size. Prey of more than 100 – 200 kg approach (Ngorongoro Crater) or even exceed (Serengeti) the amount of what a hungry hyena can eat (14.5 kg). One possible interpretation of the discrepancy between required daily meal size and those large meal sizes available at times is that hyenas are not really satisfied with the daily average (2 kg); hence, they would constantly be striving to obtain quarries of maximal reward like zebra, buffalo, or eland (*Taurotragus oryx*). If this is correct, food-procuring alone would be a sufficient selective premium to account for communal hunting. But even if larger size of prey would not offset the special social efforts of communal hunting it might under certain conditions be advantageous for other reasons: the inclusion of more prey species into the diet acts as a buffer against population fluctuations of the small and usually common prey species.

A particular problem arises when the booty that the group obtains cannot be split but is consumed by just one group member. Is it then still economic for an individual to partake in group hunting? A detailed analysis of laughing gulls stealing fish from terns (*Sterna paradisea, S. hirundo*) by Hatch (1970) provides the clue towards an answer. He found that gulls, either singly or in groups of up to eight, chased a tern that came carrying a fish to feed the brood. When pressed hard by the pursuit the tern dropped the fish, or the fish was seized from the tern's bill and eaten by only one gull. Hunting success/chase (number successful/total chases of known outcome), i.e. the tern losing its fish, rapidly approaches its peak (1.0) as group size increases (Fig. 70; see also Chap. 5 A. VII. 3.). However, hunting success in terms of fish/chase/gull continuously declines as group size increases. Why then do gulls chase terns in groups if there is no mutual advantage in gregariousness? Two points need consideration here (Hatch, manuscript). First, during chases early in the season fish/gull increases as group size rises from one to three and only declines afterwards. Observations later in the season tend to confirm the picture of Fig. 70, i.e. a continuous decrease of fish/gull with growing group size. Hence, the tactics of chasing the terns in groups becomes unprof-

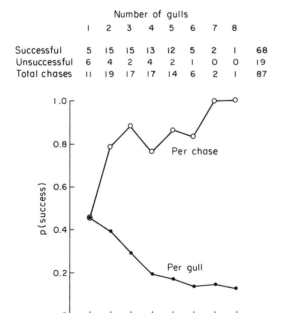

Fig. 70. Number of chases of terns by laughing gulls in groups of different sizes (*upper rows of figures*), and the probabilities of success of the chases deduced from these figures. (From Hatch, 1970)

itable for the single gull as time goes by. Possibly, time does not suffice for gulls to notice this as soon as the observer does. More important, however, appears the fact that a gull joining a chase enjoys, under all circumstances, a higher probability of getting the booty than the gull starting it, although success declines with group size. The later gulls may be more successful both because they are less fatigued than the initiator and because they have the chance to fly a shorter course and intercept the zigzagging tern. The interpretation that joiners benefit at the expense of initiators is consistent with Hatch's marked impression that the most effective stimulus for eliciting chasing is not a tern with a fish — there are always many potential targets — but a tern being chased. Joiners even flew towards chases which they could hear but not see. In conclusion, the best tactics for a gull is not to start chasing a tern, but to join one or two other gulls involved in a chase. Hence, it does not seem necessary to infer mutual advantages in this form of gregariousness, selfish individual advantage seems sufficient; also chase initiators have a definite chance, albeit a small one (Hatch, 1970). Provided gulls could assess their individual success in relation to others, it would be difficult to see why there are initiators at all if they are always at a disadvantage. One would rather expect that benefit is mutual. Benefit would be mutual, if, and only if (1) it were the same gulls that repeatedly form a party, and if (2) eventually all members succeed in obtaining fish equally often; this condition implies that the cost of initiating a chase would be equivalent to the benefits derived from joining a chase at other times. Condition (1) seems to be fulfilled since the number of gulls involved was as small as 20 pairs and there have been many hundreds of chases per season. The assumption in condition (2) can only be uncovered by extensive marking of the gulls. The two conditions just outlined would have to be expected on the basis of a model accounting for "reciprocal altruism" by Trivers (1971). It is a true alternative for explaining mutual benefits by kin selection, a condition too unlikely to have shaped the hunting habits of those gulls. Because "reciprocal altruism" only demands that altruists exchange benefits often enough so that, on average, they put in as much as they receive, the model is also able to explain equally well the mutual benefits bestowed on different species when forming mixed flocks (see p. 135). Of course, one does not know precisely how long a species would have to live for a sufficient exchange of altruistic acts to occur but birds would appear to fulfil this requirement.

Other Benefits. Apart from the benefits discussed so far there are others which may have shaped cooperative habits in the past:

The carcass may be consumed more economically. According to Andersson (1971) only pairs of long-tailed skuas (*Stercorarius longicaudus*), but not singletons, are able to tear apart and hence consume rodents completely (see also Estes and Goddard, 1967).
The carcass killed can be consumed fast enough to forestall appropriation by larger food competitors, a condition operating amongst the large African carnivores (e.g. Schaller and Lowther, 1969).
Once the prey has been brought down pack members tolerate young and sick conspecifics to join and share (Kruuk, 1972 b: hyenas, Schaller, 1972: lions), or actively

feed young by regurgitation (Kühme, 1965; Estes and Goddard, 1967; Schaller and Lowther, 1969: wild dog). Wild chimpanzees indulge, after prey capture, in a rather deliberate, long drawn-out distribution of the meat amongst the group, so that females and children which never partake in hunting get their share (Teleki, 1973).

Prey herds are believed to be less disturbed, and perhaps therefore easier to stalk, by group than by singleton hunting (Estes and Goddard, 1967).

Group members find clumps of prey by following fellow members which by their determined and speedy onward progression indicate that they have targeted prey. This has been found in various fish species (Manteifel and Radakov, 1960; Markl, 1972; Nursall, 1973; Radakov, 1973) and other animals (Chap. 3. D.). Such allelomimetic behavior is reminiscent of copying in birds (Chap. 2. C. IV. 2.) and of the suggestion that flock members exchange information about food sources (Ward, 1965; Murton, 1971; Ward and Zahavi, 1973; but see Lazarus, 1972; Gadgil, 1972). The influence of the amount of food thus located, and of other constraints, still needs to be worked out. To what extent social stimulation to feed when food is present in abundance (e.g. Welty, 1934, Uematsu, 1971, Olla and Samet, 1974) enhances the well-being of the individual, is still a matter of conjecture.

All these advantages may well be by-products of communal hunting which owes its existence to a possibly quite different selective agent. As convincingly argued by Kruuk (1972 b, p. 274 seq.), *"the* function of social hunting is the ability to overcome prey much larger or faster than the predators themselves" (italics mine). The inference is based on the correlation, within and between carnivore species, between pack size and size of quarry killed. This ecological correlation is in fact impressively neat.

When making inferences such as these one would need to establish which of several selective forces operating on a given trait is the strongest; the less effective ones which are selecting for identical properties of the species will be of no effect upon their perfection (Curio, 1973). The inference about the function of group hunting advanced by Kruuk (and less explicity by others) depends directly on particular properties of the prey can be shown to select most effectively for pack size. In this regard it may become significant that hyenas do a great deal of scavenging and that their success in appropriating a carcass from lions, for example, *also* depends strikingly on their numbers (Kruuk, 1972 b, p. 136). Future research should aim at a quantification of the various advantages in order to assess their effect as genuine selective forces.

References

Ainley, D. G.: The comfort behaviour of Adélie and other penguins. Behav. **50,** 16 – 51 (1974).

Alcock, J.: Observational learning in three species of birds. Ibis **111,** 308 – 321 (1969).

Alcock, J.: Punishment levels and the response of white-throated sparrows (*Zonotrichia albicollis*) to three kinds of artificial models and mimics. Anim. Behav. **18,** 733 – 739 (1970 a).

Alcock, J.: Punishment levels and the response of black-capped chickadees (*Parus atricapillus*) to three kinds of artificial seeds. Anim. Behav. **18,** 592 – 599 (1970 b).

Alcock, J.: Cues used in searching for food by red-winged blackbirds (*Agelaius phoeniceus*). Behav. **46,** 174 – 188 (1973 a).

Alcock, J.: The feeding response of hand-reared red-winged blackbirds (*Agelaius phoeniceus*) to a stinkbug (*Euschistus conspersus*). Am. Midland Naturalist **89,** 307 – 313 (1973 b).

Allen, J. A.: Evidence for stabilizing and apostatic selection by wild blackbirds. Nature **237,** 348 – 349 (1972).

Allen, J. A., Clarke, B.: Evidence for apostatic selection by wild passerines. Nature **220,** 501 – 502 (1968).

Allen, W. E.: Behaviour of feeding mackerel. Ecology **1,** 310 (1920 a).

Allen, W. E.: Behaviour of loon and sardines. Ecology **1,** 309 – 310 (1920 b).

Allon, N., Kochva, E.: The quantitites of venom injected into prey of different size by *Vipera palaestinae* in a single bite. J. Exptl. Zool. **188,** 71 – 76 (1974).

Altmann, M.: The flight distance in free-ranging big game. J. Wildlife Management **22,** 207 – 209 (1958).

Andersson, M.: Breeding behaviour of the long-tailed skua *Stercorarius longicaudus* (Vieillot). Ornis Scand. **2,** 35 – 54 (1971).

Andrew, R. J.: Intention movements of flight in certain passerines, and their use in systematics. Behav. **10,** 179 – 204 (1956).

Angell, T.: A study of the ferruginous hawk: adult and brood behaviour. Living Bird **9,** 225 – 241 (1970).

Apfelbach, R.: Olfactory sign stimulus for prey selection in polecats. Z. Tierpsychol. **33,** 270 – 273 (1973).

Arnold, J. W.: Feeding behaviour of a predaceous bug (Hemiptera: Nabidae). Can. J. Zool. **49,** 131 – 132 (1971).

Aschoff, J.: Jahresperiodik der Fortpflanzung bei Warmblütlern. Studium Gen. **8,** 742 – 776 (1955).

Aschoff, J.: Spontane lokomotorische Aktivität. Handbuch der Zoologie **8,** 11. Teil, Berlin: Walter de Gruyter and Co. (1962).

Aschoff, J., Wever, R.: Beginn und Ende der täglichen Aktivität freilebender Vögel. J. Orn. **103,** 2 – 27 (1962).

Athias, F.: Un cas de prédation mal dirigée de la part d'un rapace. Alauda **40,** 393 –396 (1972).

Axell, H. E.: Predation and protection at Dungeness Bird Reserve. Brit. Birds **41,** 193 – 212 (1956).

Bänsch, R.: Vergleichende Untersuchungen zur Biologie und zum Beutefangverhalten aphidovorer Coccinelliden, Chrysopiden und Syrphiden. Zool. Jb. Syst. **91,** 271 – 340 (1964).

Baerends, G. P.: Specialications in organs and movements with a releasing function. In: Physiological Mechanisms in Animal Behaviour. Symp. Soc. Exptl. Biol. **4,** 337 – 360. Cambr. Univ. Press 1950.

Baker, M. C.: Foraging behaviour of black-bellied plovers (*Pluvialis squatarola*). Ecology **55,** 162 – 167 (1974).

Bakus, G. J.: Defensive mechanisms and ecology of some tropical holothurians. Marine Biol. **2,** 23 – 32 (1968).

Balan, J., Gerber, N. N.: Attraction and killing of the nematode *Panagrellus redivivus* by the predaceous fungus *Arthrobotrys dactyloides.* Nematologica **18,** 163 – 173 (1972).

Banks, C. J.: The behaviour of individual coccinellid larvae on plants. Brit. J. Anim. Behav. **5,** 12 – 24 (1957).

Banner, A.: Use of sound in predation by young lemon sharks, *Negaprion brevirostris* (Poey). Bull. Marine Sci. **22,** 251 – 283 (1972).

Bartholomew, G. A.: The fishing activities of double-crested cormorants on San Francisco Bay. Condor **44,** 13 – 21 (1942).

Bauer, K. M., Glutz von Blotzheim, U. N.: Handbuch der Vögel Mitteleuropas. **1,** Gaviiformes – Phoenicopteriformes. Frankfurt: Akad. Verlagsges. 1966.

Bauer, K. M., Glutz von Blotzheim, U. N.: Handbuch der Vögel Mitteleuropas. **2,** Anseriformes (Part 1). Frankfurt: Akad. Verlagsges. 1968.

Beal, F. E. L.: Food habits of the swallows, a family of valuable native birds. U. S. Dept. Agr. Bull. **619,** 1 – 28 (1918).

Beattie, A. J., Breedlove, D. E., Ehrlich, P. R.: The ecology of the pollinators and predators of *Frasera speciosa.* Ecology **54,** 81 – 91 (1973).

Benzie, V. L.: Some aspects of the anti-predator responses of two species of sticklebacks. Ph. D. thesis, Oxford Univ. Oxford 1965.

Berlyne, D. E.: Conflict, Arousal, and Curiosity. New York, Toronto, London: McGraw-Hill 1960.

Berndt, R.: Zur Bestandsentwicklung der Greifvögel (Falconiformes) im Drömling. Beitr. Vogelk. **16,** 1 – 12 (1970).

Berry, P. Y.: The diet of some Singapore Anura (Amphibia). Proc. Zool. Soc. London. **144,** 163 – 174 (1965).

Berry, P. Y.: The food of the giant toad *Bufo asper.* Zool. J. Linn. Soc. **49,** 61 – 68 (1970).

Berthold, P., Berthold, H.: Jahreszeitliche Änderungen der Nahrungspräferenz und deren Bedeutung bei einem Zugvogel. Naturwiss. **60,** 391 – 392 (1973).

Bethel, W. M.: Altered behavior leading to selective predation of amphipods infected with acanthocephalans, with special reference to *Polymorphus paradoxus.* Ph. D. thesis, Univ. Alberta, Edmonton (1972).

Beukema, J. J.: Predation by the three-spined stickleback (*Gasterosteus aculeatus* L.): The influence of hunger and experience. Behav. **31,** 1 – 126 (1968).

Bigelow, H. B., Schroeder, W. C.: Sharks. Mem. Sears Found. Marine Res. **1,** 59 – 546 (1948).

Black, J. H.: A possible stimulus for the formation of some aggregations in tadpoles of *Scaphiopus bombifrons.* Proc. Oklahoma Acad. Sci. **49,** 13 – 14 (1970).

Blest, A. D.: Longevity, palatability and natural selection in five species of New World saturniid moths. Nature **197,** 1183 – 1186 (1963).

Boer, den, M. H.: A colour-polymorphism in caterpillars of *Bupalus piniarius* (L.). Neth. J. Zool. **21,** 61 – 116 (1971).

Boletzky, S. v.: A note on aerial prey capture by *Sepia officinalis* (Mollusca, Cephalopoda). Vie Milieu **23,** Ser. Biol. Marine, 133 – 140 (1972).

Boulet, P. C.: Contribution à l'étude experimentale de la perception visuelle du mouvement chez la perche et la seiche. Mem. Museum Natl. Hist. Nat. (Paris), Ser. A. Zool. **17,** (1958).

Bowman, R. J: Morphological differentiation and adaptation in the Galapagos Finches. Berkeley, Los Angeles: Univ. Calif. Press 1961.

Bräuning, C., Lichtner, J.-C.: Gemeinsame Jagd zweier Merline. Vogelwelt **91,** 32 (1970).

212

Bragg, A. N.: Experimental observations on the feeding of spadefoot tadpoles. Southwestern Naturalist **5**, 201 – 207 (1960).

Bragg, A. N.: *Saprolegnia* on tadpoles again in Oklahoma. Southwestern Naturalist **7**, 79 – 80 (1962).

Bragg, A. N.: Mass movements resulting in aggregations of tadpoles of the plains spadefoot, some of them in response to light and temperature (Amphibia: Salientia). Wasman J. Biol. **22**, 299 – 305 (1964).

Bragg, A. N.: Longevity of the tadpole stage in the plains spadefoot (Amphibia: Salientia). Wasman J. Biol. **24**, 71 – 73 (1966).

Bragg, A. N., Bragg, W. N.: Variations in the mouth part of *Scaphiopus* (*Spea*) *bombifrons* Cope (Amphibia: Salientia). Southwestern Naturalist **3**, 55 – 69 (1959).

Brandl, Z.: Laboratory culture of cyclopoid copepods on a definite food. Vestn. Cesk. Spolecnosti Zool. **37**, 81 – 88 (1973).

Braum, E.: Die ersten Beutefanghandlungen junger Blaufelchen (*Coregonus wartmanni* Bloch) und Hechte (*Esox lucius* L.). Z. Tierpsychol. **20**, 247 – 266 (1963).

Breder, C. M., Jr.: On the survival value of fish schools. Zoologica **52**, 25 – 40 (1967).

Brockhusen, F. von, Curio, E.: Die innerartliche Variabilität der Beutewahl beuteerfahrungsloser Anolis. Experientia **31**, 45 – 46 (1975).

Brosset, A.: La vie sociale des oiseaux dans une forêt équatoriale du Gabon. Extr. rev. Biol. Gabon **5**, 26 – 69 (1969).

Brosset, A.: Recherches sur la biologie des Pycnonotidés du Gabon. Biol. Gabonica **7**, 423 – 460 (1971).

Brown, L.: Selection in a population of house mice containing mutant individuals. J. Mammal, **46**, 461 – 465 (1965).

Brown, L.: Eagles. London, New York: A. Barker Ltd. and Arco Publ. Co. Inc. 1970.

Brown, L., Amadon, D.: Eagles, Hawks, and Falcons of the World. Vols. 1, 2. Feltham: Hamlyn House 1968.

Bruns, E. H.: Winter predation of golden eagles and coyotes on pronghorn antelopes. Can. Field-Natur. **84**, 301 – 403 (1970).

Bryan, J. E.: Feeding history, parental stock, and food selection in rainbow trout. Behav. **45**, 123 – 153 (1973).

Bryan, J. E., Larkin, P. A.: Food specialization by individual trout. J. Fisheries Res. Board Can. **29**, 1615 – 1624 (1972).

Buckley, F. G., Buckley, P. A.: The breeding ecology of royal terns *Sterna* (*Thalasseus*) *maxima maxima*. Ibis **114**, 344 – 359 (1972).

Buckley, F. G., Buckley, P. A.: Comparative feeding ecology of wintering adult and juvenile royal terns (Aves: Laridae, Sterninae). Ecology **55**, 1053 – 1063 (1974).

Buckley, P. A., Buckley, F. G.: Tongue-flicking by a feeding snowy egret. Auk **85**, 678 (1968).

Bullis, H. R., Jr.: Observations on the feeding behavior of white-tip sharks on schooling fishes. Ecology **42**, 194 – 195 (1961).

Burghardt, G. M.: Effects of prey size and movements on the feeding behaviour of the lizards *Anolis carolinensis* and *Eumeces fasciatus*. Copeia **3**, 576 – 578 (1964).

Burghardt, G. M.: Stimulus control of the prey attack response in naive garter snakes. Psychon. Sci. **4**, 37 – 38 (1966).

Burghardt, G. M.: Chemical-cue preferences of inexperienced snakes: comparative aspects. Science **157**, 718 – 721 (1967).

Burghardt, G. M.: Effects of early experience on food preference in chicks. Psychon. Sci. **14**, 7 – 8 (1969).

Burghardt, G. M.: Chemical perception in reptiles. In: Communication by Chemical Signals. J. W. Johnson, D. G. Moulton and A. Turk. (eds.). New York: Appleton-Century-Crofts 1970 a.

Burghardt, G. M.: Intraspecific geographical variation in chemical food cue preferences of newborn garter snakes (*Thamnophis sirtalis*). Behav. **36**, 246 – 257 (1970 b).

Burghardt, G. M.: Chemical release of prey attack: Extension to naive newly hatched lizards, *Eumeces fasciatus.* Copeia **1**, 178 – 181 (1973).

Burghardt, G. M.: Chemical prey preference polymorphism in newborn garter snakes, *Thamnophis sirtalis.* Behav. **52**, 202 – 225 (1975).

Burghardt, G. M., Abeshaheen, J. P.: Responses to chemical stimuli of prey in newly hatched snakes of the genus *Elaphe.* Anim. Behav. **19**, 486 – 489 (1971).

Burghardt, G. M., Hess, E. H.: Food imprinting in the snapping turtle, *Chelydra serpentina.* Science **151**, 108 – 109 (1966).

Burghardt, G. M., Hess, E. H.: Factors influencing the chemical release of prey attack in newborn snakes. Comp. Physiol. Psychol. **66**, 289 – 295 (1968).

Busnel, R. G. (ed.): Acoustic Behaviour of Animals. Amsterdam, London, New York: Elsevier Publ. Co. 1963.

Büttner, K.: Untersuchungen über den Einfluß des Beutetiers auf den Erbeutungsvorgang bei der Waldameise, *Formica polyctena* Foerster (Hymen., Formicidae). Z. Angew. Entomol. **74**, 177 – 196 (1973).

Cade, T. J.: Ecology of the peregrine and gyrfalcon populations in Alaska. Univ. Calif. Publ. Zool. **63**, 151 – 290 (1960).

Cade, T. J.: Ecological and behavioral aspects of predation by the northern shrike. Living Bird **6**, 43 – 86 (1967).

Campbell, H. W.: Prey selection in naive *Elaphe obsoleta* (Squamata: Serpentes) –A reappraisal. Psychon. Sci. **21**, 300 – 301 (1970).

Canella, M. F.: Experimental researches on monocular vision of *Chamaeleo* and *Melopsittacus.* Proc. 16th. Intern. Congr. Zool. **2**, 18 (1963).

Carpenter, G. D. H.: Observations and experiments in Africa by the late C. F. M. Swynnerton on wild birds eating butterflies and the preference shown. Proc. Linnean Soc. London **154**, Part I, 10 – 46 (1942).

Carriker, M. L.: Observation on the penetration of tightly closing bivalves by *Busycon* and other predators. Ecology **32**, 73 – 83 (1951).

Carriker, M. R., van Zandt, D.: Predatory behavior of a shell-boring muricid gastropod. Behav. Marine Animal. **1**, 157 – 244 (1972).

Chandler, A. E. F.: Locomotory behaviour of first instar larvae of aphidophagous Syrphidae (Diptera) after contact with aphids. Anim. Behav. **17**, 673 – 678 (1969).

Chew, K. K.: Study of food preference and rate of feeding of Japanese oyster drill, *Ocinebra japonica* (Dunker). U. S. Fish Wildlife Serv., Spec. Sci. Rep. Fisheries No. 365 (1960).

Clarke, B.: Balanced polymorphism and the diversity of sympatric species. In: Taxonomy and Geography. D. Nichols. (ed.), Syst. Assoc. Publ. **4**, 47 – 70 (1962).

Clarke, B.: The evidence for apostatic selection. Heredity **24**, 347 – 352 (1969).

Clarke, B., O'Donald, P.: Frequency-dependent selection. Heredity **19**, 201 – 206 (1964).

Clarke, T. A., Flechsig, A. O., Grigg, R. W.: Ecological studies during project Sealab. II. Science **157**, 1381 – 1389 (1967).

Cloarec, A.: Etude déscriptive et experimentale du comportement des capture de *Ranatra linearis* au cours de son ontogenése. Behavior **35**, 83 – 113 (1969).

Cloudsley-Thompson, J. L.: The classification and study of animals by feeding habits. I. E. F. N. Biology of Nutrition **18**, 439 – 470. R. N. Fiennes (ed.), Oxford, New York: Pergamon Press 1972.

Coble, D. W.: Influence of appearance of prey and satiation of predator on food selection by Northern Pike (*Esox lucius*). J. Fisheries Res. Board Can. **30**, 317 – 320 (1973).

Connell, J. H.: Effects of competition, predation by *Thais lapillus* and other factors on natural populations of the barnacle *Balanus balanoides.* Ecol. Monogr. **31**, 61 – 104 (1961).

Cook, A. D., Atsatt, P., Simon, C. A.: Doves and dove weed: Multiple defense against avian predation. BioScience **21**, 277 – 281 (1971).

Cooke, A. S.: Selective predation by newts on frog tadpoles treated with DDT. Nature **229**, 275 – 276 (1971).

214

Coppinger, R. P.: The effect of experience and novelty on avian feeding behaviour with references to the evolution of warning coloration in butterflies. Part I: Reactions of wild-caught adult blue jays to novel insects. Behav. **35**, 45 – 60 (1969).

Coppinger, R. P.: The effect of experience and novelty on avian feeding behaviour with reference to the evolution of warning coloration in butterflies. Part II: Reactions of naive birds to novel insects. Am. Naturalist **104**, 323 – 335 (1970).

Cott, H. B.: Adaptive coloration in animals. London: Methuen and Co. 1957.

Craighead, J. J., Craighead, F. C.: Hawks, Owls and Wildlife. Harrisburg: The Stackpole Co., 1956.

Crisler, L.: Observations of wolves hunting caribou. J. Mammal. **37**, 337 – 346 (1956).

Crisp, M.: Studies on the behaviour of *Nassarius obsoletus* (Say) (Mollusca, Gastropoda). Biol. Bull. **136**, 355 – 373 (1969).

Crome, W.: Arachnida. In: Exkursionsfauna, Wirbellose I, 289 – 361. E. Stresemann (ed.), Berlin: Volk und Wissen 1957.

Crowcroft, P.: The daily cycle of activity in British shrews. Proc. Zool. Soc. London **123**, 715 – 729 (1954).

Croze, H.: Searching image in carrion crows. Z. Tierpsychol. Beih. **5**, 1 – 85 (1970).

Curio, E.: Beobachtungen am Halbringschnäpper, *Ficedula semitorquata,* im mazedonischen Brutgebiet. J. Orn. **100**, 176 – 209 (1959 a).

Curio, E.: Beobachtungen an einem Rötelammerpaar (*Pipilo erythrophthalmus*) in Gefangenschaft. Gefied. Welt, 1 – 3 and 25 – 29 (1959 b).

Curio, E.: Verhaltensstudien am Trauerschnäpper. Z. Tierpsychol. Beih. **3**, 1 – 118 (1959 c).

Curio, E.: Probleme des Feinderkennens bei Vögeln. Proc. 13th Intern. Orn. Congr., 106 – 239 (1963).

Curio, E.: Die Schutzanpassungen dreier Raupen eines Schwärmers (Lepidopt., Sphingidae) auf Galapagos. Zool. Jb. Syst. **92**, 487 – 522 (1965 a).

Curio, E.: Galapagos – Prüffeld der Evolutionsforschung. Umschau **65**, 562 – 567 (1965 b).

Curio, E.: Ein Falter mit „falschem Kopf". Natur u. Museum **95**, 43 – 46 (1965 c).

Curio, E.: Funktionsweise und Stammesgeschichte des Flugfeinderkennens einiger Darwinfinken (Geospizinae). Z. Tierpsychol. **26**, 394 – 487 (1969).

Curio, E.: Die Messung des Selektionswertes einer Verhaltensweise. Verhandl. Deut. Zool. Ges., 64. Tagung, 348 – 352. Stuttgart: Gustav Fischer Verlag 1970 a.

Curio, E.: Die Selektion dreier Raupenformen eines Schwärmers (Lepidopt., Sphingidae) durch einen Anolis (Rept., Iguanidae). Z. Tierpsychol. **27**, 899 – 914 (1970 b).

Curio, E.: Towards a methodology of teleonomy. Experientia **29**, 1045 – 1058 (1973).

Curio, E.: The functional organization of anti-predator behaviour in the Pied Flycatcher: A study of avian visual perception. Anim. Behav. **23**, 1 – 115 (1975).

Curio, E., Kramer, P.: Vom Mangrovefinken (*Cactospiza heliobates* Snodgrass und Heller). Z. Tierpsychol. **21**, 223 – 234 (1964).

Daly, J. W., Witkop, B.: Chemistry and Pharmacology of Frog Venoms. In: Venomous Animals and their Venoms. W. Bücherl and E. E. Buckley (eds.), **2**, 497 – 519. New York and London: Academic Press 1971.

Davis, J.: Comparative foraging behaviour of the spotted and brown towhees. Auk **74**, 129 – 166 (1957).

Dawkins, M.: Perceptual changes in chicks: Another look at the "search image" concept. Anim. Behav. **19**, 566 – 574 (1971 a).

Dawkins, M.: Shifts of "attention" in chicks during feeding. Anim. Behav. **19**, 575 – 582 (1971 b).

Dethier, V. G.: Feeding and drinking behavior of invertebrates. Handbook of Physiol., 79 – 96. C. F. Code (ed.), Am. Physiol. Soc. 1967.

Dethier, V. G., Bodenstein, D.: Hunger in the blowfly. Z. Tierpsychol. **15**, 129 – 140 (1958).

Dice, L. R.: Effectiveness of selection by owls of deer-mice (*Peromyscus maniculatus*) which contrast in colour with their background. Contrib. Lab. Vert. Biol. Univ. Mich., Ann Arbor, **34**, 1 – 20 (1947).

Dill, L. M.: An avoidance learning submodel for a general predation model. Oecologia **13**, 291 – 312 (1973).

Dinsmore, J. J.: Sooty tern behavior. Bull. Florida State Mus., Biol. Sci. **16**, 129 – 179 (1972).

Dixon, A. F. G.: An experimental study of the searching behaviour of the predatory coccinellid beetle *Adalia decempunctata* (L.). J. Anim. Ecol. **28**, 259 – 281 (1959).

Drees, O.: Untersuchungen über die angeborenen Verhaltensweisen bei Springspinnen (Salticidae). Z. Tierpsychol. **9**, 169 – 207 (1952).

Drinnan, R. E.: The winter feeding of the oystercatcher (*Haematopus ostralegus*) on the edible cockle. J. Animal Ecol. **26**, 439 – 469 (1957).

Dunn, E. K.: Robbing behaviour of roseate terns. Auk **90**, 641 – 651 (1973 a).

Dunn, E. K.: Changes in fishing ability of terns associated with windspeed and sea surface conditions. Nature **244**, 520 – 521 (1973 b).

Dunstan, T. C.: Feeding activities of ospreys in Minnesota. Wils. Bull. **86**, 74 – 76 (1974).

Eastman, R.: The Kingfisher. London: Collins, 1969.

Eaton, R. L.: The predatory sequence, with emphasis on killing behaviour and its ontogeny, in the cheetah (*Acinonyx jubatus* Schreber). Z. Tierpsychol. **27**, 492 – 504 (1970).

Edmunds, M.: Defence in animals. Harlow, Essex: Longman Group Ltd. 1974.

Edwards, J. S.: Arthropods as predators. In: Viewpoints in Biology. Vol. **2**, 85 – 114. J. D. Carthy and C. L. Duddington (eds.), London: Butterworth Inc. 1963.

Ehrlich, P. R., Ehrlich, A. H.: Coevolution: Heterotypic schooling in Caribbean Reef fishes. Am. Naturalist **107**, 157 – 160 (1973).

Ehrman, L.: Simulation of the mating advantage of rare *Drosophila* males. Science **167**, 905 – 906 (1970).

Eibl-Eibesfeldt, I.: Nahrungserwerb und Beuteschema der Erdkröte (*Bufo bufo* L.). Behav. **4**, 1 – 35 (1952).

Eibl-Eibesfeldt, I.: Über Symbiosen, Parasitismus und andere besondere zwischenartliche Beziehungen tropischer Meeresfische. Z. Tierpsychol. **12**, 203 – 219 (1955 a).

Eibl-Eibesfeldt, I.: Zur Biologie des Iltis (*Putorius putorius* L.). Verhandl. Deut. Zool. Ges., 304 – 314 (1955 b).

Eibl-Eibesfeldt, I.: *Putorius putorius* (L.). Beutefang I (Töten von Wanderratten). Inst. Wiss. Film, Göttingen 1958 a.

Eibl-Eibesfeldt, I.: *Putorius putorius* (L.). Beutefang IV (Töten von Kreuzottern). Inst. Wiss. Film, Göttingen 1958 b.

Eibl-Eibesfeldt, I.: Das Verhalten der Nager. Handb. Zool. **8**, 10 – 12, 1 – 87 (1958).

Eibl-Eibesfeldt, I.: Die Verhaltensentwicklung des Krallenfrosches (*Xenopus laevis*) und des Scheibenzünglers (*Discoglossus pictus*) unter besonderer Berücksichtigung der Beutefanghandlungen. Z. Tierpsychol. **19**, 385 – 393 (1962 a).

Eibl-Eibesfeldt, I.: Freiwasserbeobachtungen zur Deutung des Schwarmverhaltens verschiedener Fische. Z. Tierpsychol. **19**, 165 – 182 (1962 b).

Eibl-Eibesfeldt, I.: Grundriß der vergleichenden Verhaltensforschung. München: Piper 1967.

Eibl-Eibesfeldt, I., Hass, H.: Erfahrungen mit Haien. Z. Tierpsychol. **16**, 739 – 746 (1959).

Eibl-Eibesfeldt, I., Sielmann, H.: Beobachtungen am Spechtfinken *Cactospiza pallida* (Sclater und Salvin). J. Orn. **103**, 92 – 101 (1962).

Eisner, R., Alsop, R., Ettershank, G.: Adhesiveness of spider silk. Science **146**, 1058 – 1061 (1964).

Eisner, R., Silberglied, R. E., Aneshansley, D., Carrel, J. E., Howland, H. C.: Ultraviolet video-viewing: The television camera as an insect eye. Science **166**, 1172 – 1174 (1969).

216

Eloff, F. C.: On the predatory habits of lions and hyaenas. Koedoe **7**, 105 – 112 (1964).

Emlen, J. M.: Ecology: An Evolutionary Approach. Reading, Mass.: Addison-Wesley 1973.

Emlen, J. T., Miller, D. E., Evans, R. M., Thompson, D. H.: Predator-induced parental neglect in a ring-billed gull colony. Auk **83**, 677 – 679 (1966).

Erkinaro, E.: Seasonal changes in the phase position of circadian activity rhythms in some voles, and their endogenous component. Aquilo Ser. Zool. **13**, 87 – 91 (1972).

Erkinaro, E.: Short-term rhythm of locomotor activity within the 24 h period in the Norwegian lemming, *Lemmus lemmus,* and water vole, *Arvicola terrestris.* Aquilo Ser. Zool. **14**, 56 – 58 (1973).

Errington, P. L.: Predation and vertebrate populations. Quart. Rev. Biol. **21**, 145 – 177 and 221 – 245 (1946).

Errington, P. L.: Of predation and life. Ames: Iowa State Univ. Press 1967.

Estes, R. D., Goddard, J.: Prey selection and hunting behaviour of the African wild dog. J. Wildlife Management **31**, 52 – 70 (1967).

Etienne, A. S.: Analyse der schlagauslösenden Bewegungsparameter einer punktförmigen Beuteattrappe bei der *Aeschna*larve. Z. vergl. Physiol. **64**, 71 – 110 (1969).

Etienne, A. S.: The behaviour of the dragonfly larva *Aeschna cyanea* M. after a short presentation of a prey. Anim. Behav. **20**, 724 – 731 (1972).

Etienne, A. S.: Searching behaviour towards a disappearing prey in the domestic chick as affected by preliminary experience. Anim. Behav. **21**, 749 – 758 (1973).

Eutermoser, A.: Schlagen Beizfalken bevorzugt kranke Krähen? Vogelwelt **82**, 101 – 104 (1961).

Evans, W. E., Bastian, J.: Marine mammals: social and ecological problems. In: Biology of Marine Mammals. H. T. Andersen (ed.), New York: Academic Press 1969.

Ewer, R. F., Wemmer, C.: The behaviour in captivity of the African civet, *Civettictis civetta* (Schreber). Z. Tierpsychol. **34**, 359 – 394 (1974).

Ewert, J. P.: Der Einfluß von Zwischenhirndefekten auf die Visuomotorik im Beute- und Fluchtverhalten der Erdkröte (*Bufo bufo* L.). Z. vergl. Physiol. **61**, 41 – 70 (1968).

Ewert, J. P.: Aufnahme und Verarbeitung visueller Informationen im Beutefang- und Fluchtverhalten der Erdkröte *Bufo bufo* (L.). Verhandl. Deut. Zool. Ges. 218 – 226. Stuttgart: Fischer 1970.

Ewert, J. P., Börchers, H. W.: Reaktionscharakteristik von Neuronen aus dem Tectum opticum und Sub-Tectum der Erdkröte (*Bufo bufo* L.). Z. vergl. Physiol. **68**, 84 – 110 (1970).

Fallet, M.: Über Bodenvögel und ihre terricolen Beutetiere. Technik der Nahrungssuche – Populationsdynamik. Zool. Anz. **168**, 187 – 212 (1962).

Farmer, G. J., Beamish, F. W.: Sea lamprey (*Petromyzon marinus*) predation on freshwater teleosts. J. Fisheries Res. Board Can. **30**, 601 – 605 (1973).

Feder, H. M.: Cleaning symbioses in the marine environment. In: Symbiosis, Vol. **1**, 327 – 380. S. M. Henry (ed.), New York: Academic Press 1966.

Federighi, H.: Studies on the oyster drill (*Urosalpinx cinerea,* Say). Bull. U. S. Bur. Fish **47**, 85 – 115 (1931).

Fink, B. D.: Observations of porpoise predation on a school of Pacific sardines. Calif. Fish Game **45**, 216 – 217 (1959).

Fischer, W.: Die Seeadler. Wittenberg-Lutherstadt: Ziemsen 1970.

Fisher, J., Peterson, R. T.: The World of Birds. London: Macdonald 1964.

Fisher, R. C.: Aspects of the physiology of endoparasitic Hymenoptera. Biol. Rev. **46**, 234 – 278 (1971).

Fleschner, C. A.: Studies on searching capacity of the larvae of three predators of the citrus red mite. Hilgardia **20**, 233 – 265 (1950).

Fletcher, H. J.: The delayed response problem. In: Behavior of Nonhuman Primates. Vol. **1**, 129 – 165. A. M. Schrier, H. F. Harlow, F. Stollnitz (eds.), New York, London: Academic Press 1965.

Ford, E. B.: Ecological Genetics. London: Chapman and Hall 1971.

Foster, N. R.: Behaviour, development, and early life history of the Asian needle-fish, *Cenentodon cancila*. Proc. Natl. Acad. Sci. U. S. **125**, 77 – 88 (1973).

Foster, N. R., Scheir, A., Cairns, J., Jr.: Effects of ABS on feeding behaviour of flag-fish, *Jordanella floridae*. Trans. Am. Fisheries Soc. **95**, 109 – 110 (1966).

Fox, M. W.: Socio-ecological implications of individual differences in wolf litters: A developmental and evolutionary perspective. Behav. **41**, 298 – 313 (1972).

Frame, G. W.: Black bear predation on Salmon at Olsen Creek, Alaska. Z. Tierpsychol. **35**, 23 – 88 (1974).

Freisling, J.: Studien zur Biologie und Psychologie der Wechselkröte (*Bufo viridis* Laur). Österr. Zool. Z. **1**, 383 – 440 (1948).

Fricke, H. W.: Zwischenartliche Beziehungen der tropischen Meerbarsche *Pseudupeneus barberinus* und *Pseudupeneus macronema* mit einigen anderen marinen Fischen. Natur Museum **100**, 71 – 80 (1970).

Fricke, H. W.: Fische als Feinde tropischer Seeigel. Marine Biol. **9**, 328 – 338 (1971).

Fricke, H. W.: Der Einfluß des Lichtes auf Körperfärbung und Dämmerungsverhalten des Korallenfisches *Chaetodon melanotus*. Marine Biol. **22**, 251 – 262 (1973).

Friedmann, H.: Avian symbiosis. In: Symbiosis, Vol. II. S. M. Henry (ed.), New York: Academic Press 1967.

Frings, H., Frings, M., Cox, B., Preissner, L.: Auditory and visual mechanisms in food-finding behaviour of the herring gull. Wils. Bull. **67**, 155 – 170 (1955).

Frisch, K. v.: Bestrafte Gefräßigkeit. Z. Tierpsychol. **16**, 647 – 650 (1959).

Frisch, K. v.: Tanzsprache und Orientierung der Bienen. Berlin, Heidelberg, New York: Springer 1965.

Frisch, O. v.: Zur Biologie des Zwergchamäleons (*Microsaurus pumilus*). Z. Tierpsychol. **19**, 276 – 289 (1962).

Fryer, G., Iles, T. D.: The cichlid fishes of the great lakes of Africa. Edinburgh: Oliver and Boyd 1972.

Fuchs, J. L., Burghardt, G. M.: Effects of early feeding experience on the responses of garter snakes to food chemicals. Learn. Motiv. **2**, 271 – 279 (1971).

Fuldner, D., Wolf, H.: Staphyliniden-Larven beeinflussen chemisch das Orientierungsverhalten ihrer Konkurrenten. Naturwissenschaften **58**, 418 – 419 (1971).

Gadgil, M.: The function of communal roosts: relevance of mixed roosts. Ibis **114**, 531 – 533 (1972).

Galbraith, M. G., Jr.: Size-selective predation on *Daphnia* by rainbow trout and yellow perch. Trans. Am. Fisheries Soc. **96**, 1 – 10 (1967).

Galef, B. H., Sherry, D. F.: Mother's milk: A medium for transmission of cues reflecting the flavor of mother's diet. J. Comp. Physiol. Psychol. **83**, 374 – 378 (1973).

Galtsoff, P. S., Prytherch, H. F., Engle, J. B.: Natural history and methods of controlling the common oyster drills (*Ursosalpinx cinerea* Say and *Eupleura caudata* Say). U. S. Bur. Fisheries, Fisheries Circ. **25**, 1 – 24 (1937).

Gandolfi, G., Mainardi, D., Rossi, A. C.: La reazione di paura e lo svantaggio individuale dei pesci allarmisti (esperimenti con modelli). Zoologia **102**, 8 – 14 (1968).

Gardner Tugendhat, B.: Hunger and sequential responses in the hunting behaviour of Salticid spiders. J. Comp. Physiol. Psychol. **58**, 167 – 173 (1964).

Gaston, A. J.: The ecology and behaviour of the long-tailed tit. Ibis **115**, 332 – 351 (1973).

Gennaro, J. F., Leopold, R. S., Merriam, T. W.: Observations on the actual quantity of venom introduced by several species of crotalid snakes in their bite. Anat. Record **139**, 303 (1961).

Gibb, J. A.: Feeding ecology of tits, with notes on treecreeper and goldcrest. Ibis **96**, 513 – 543 (1954).

Gibb, J. A.: Predation by tits and squirrels on the eucosmid *Ernarmonia conicolana* (Heyl.). J. Anim. Ecol. **27**, 375 – 396 (1958).

218

Gibb, J. A.: L. Tinbergen's hypothesis of the role of specific search images. Ibis **104**, 106 – 111 (1962).

Gibb, J. A.: Tit predation and the abundance of *Ernarmonia conicolona* (Heyl.) on Weeting Heath, Norfolk, 1962 – 1963. J. Anim. Ecol. **35**, 43 – 53 (1966).

Gibson, W.: Effect of a predator on the sleep of a prey. Psychology **6**, 231 (1970).

Gibson, D. O.: Batesian mimicry without distastefulness? Nature **250**, 77 – 79 (1974).

Giese, A. C., Alden, R. H.: Cannibalism and giant formation in *Stylonychia*. J. Exptl. Zool. **78**, 117 (1938).

Gilbert, J. J.: The adaptive significance of polymorphism in the rotifer *Asplanchna*. Humps in males and females. Oecologia **13**, 135 – 146 (1973).

Glass, B.: In discussion of The evolution of mimicry; a problem in ecology and genetics. (Sheppard, P. M.). Cold Spring Harbor Symp. Quant. Biol. **24**, 140 (1959).

Glutz v. Blotzheim, U. N., Bauer, K. M., Bezzel, E.: Handbuch der Vögel Mitteleuropas. Vol. 4. Falconiformes. Frankfurt: Akad. Verlagsges. 1971.

Godfrey, G. K.: A field study of the activity of the mole (*Talpa europaea*). Ecology **36**, 678 – 685 (1955).

Godfrey, G. K., Crowcroft, P.: The Life of the Mole. London: Museum Press 1960.

Goodyear, C. P.: Learned orientation in the predator avoidance behaviour of mosquito fish, *Gambusia affinis*. Behav. **45**, 191 – 220 (1973).

Gordon, S.: The Golden Eagle. London: Collins 1955.

Goslow, G. E., Jr.: The attack and strike of some North American raptors. Auk **88**, 815 – 827 (1971).

Goss-Custard, J. D.: The response of redshank (*Tringa totanus* L.) to spatial variations in the density of their prey. J. Anim. Ecol. **39**, 91 – 113 (1970 a).

Goss-Custard, J. D.: Factors affecting the diet and feeding rates of the redshank (*Tringa totanus*). Symp. Brit. Ecol. Soc. **10**, 101 – 110 (1970 b).

Gossow, H.: Vergleichende Verhaltensstudien an Marderartigen. I. Über Lautäußerungen und zum Beuteverhalten. Z. Tierpsychol. **27**, 405 – 480 (1970).

Grant, P. R.: Interactive behaviour of puffins (*Fratercula arctica* L.) and skuas (*Stercorarius parasiticus* L.). Behav. **40**, 263 – 281 (1971).

Greenhall, A. M., Schmidt, U., Lopez-Forment, W.: Attacking behaviour of the vampire bat, *Desmodus rotundus*, under field conditions in Mexico. Biotropica **3**, 136 – 141 (1971).

Grönlund, S., Itämies, J., Mikkola, H.: On the food and feeding habits of the great grey shrike *Lanius excubitor* in Finland. Ornis Fenn. **47**, 167 – 171 (1970).

Grzimek, B.: Grzimeks Tierleben. Vol. XIII. Säugetiere 4. Zürich: Kindler 1968.

Grzimek, B.: Grzimeks Tierleben. Vol. IV. Fische I. Zürich: Kindler 1970.

Grzimek, B.: Grzimeks Tierleben. Vol. V. Fische 2 und Lurche. Zürich: Kindler 1970.

Grzimek, B.: Grzimeks Tierleben. Vol. VI. Kriechtiere. Zürich: Kindler 1971.

Grzimek, B.: Grzimeks Tierleben. Vol. XII. Säugetiere 3. Zürich: Kindler 1972.

Guggisberg, C. A. W.: Crocodiles. Their Natural History, Folklore and Conservation. Newton Abbot: David and Charles Ltd. 1972.

Gundlach, H.: Brutfürsorge, Brutpflege, Verhaltensontogenese und Tagesperiodik beim Europäischen Wildschwein (*Sus scrofa* L.). Z. Tierpsychol. **25**, 955 – 995 (1968).

Gwinner, E.: Über den Einfluß des Hungers und anderer Faktoren auf die Versteck-Aktivität des Kolkraben (*Corvus corax*). Vogelwarte **23**, 1 – 4 (1965).

Gwinner, E.: Adaptive function of circannual rhythms in warblers. Proc. 15th. Intern. Ornithol. Congr., Leiden 218 – 236 (1972).

Haartman, L. v.: Was reizt den Trauerfliegenschnäpper (*Muscicapa hypoleuca*) zu füttern? Vogelwarte **4**, 157 – 164 (1953).

Haartman, L. v.: Der Trauerfliegenschnäpper. III. Die Nahrungsbiologie. Acta Zool. Fennica **83**, 6 – 91 (1954).

Hailman, J. P.: The Galapagos swallow-tailed gull is nocturnal. Wils. Bull. **76**, 347 – 354 (1964).

Hall, J. R.: Intraspecific trail-following in the marsh periwinkle *Littorina irrorata* Say. Veliger **16**, 72 – 75 (1973).

Hall, K. R. L.: Tool-using performances as indicators of behavioural adaptability. Current Anthropol. **4**, 479 – 494 (1963).

Hamerstrom, F.: An eagle to the sky. Ames Iowa: Iowa State Univ. Press 1970.

Hamilton, W. D.: Geometry for the selfish herd. J. Theoret. Biol. **31**, 295 – 311 (1971).

Hamilton, W. J. III.: Life's colour code. New York: McGraw Hill Co. 1973.

Hamilton, W. J., Hamilton, M. E.: Breeding characteristics of yellow-billed cuckoos in Arizona. Proc. Calif. Acad. Sci. **32**, 405 – 432 (1965).

Harding, R. S. O.: Predation by a troop of olive baboons (*Papio anubis*). Paper delivered 4th. Intern.Primatol. Congr., Portland, Oregon, August 18th, 1972.

Harris, M. P.: The food of some *Larus* gulls. Ibis **107**, 43 – 53 (1965).

Hartley, P. H. T.: Feeding habits. In: A new Dictionary of birds. A. Landsborough-Thomson (ed.), London, Edinburgh: Nelson, 1964.

Harvey, P. H., Jordan, C. A., Allen, J. A.: Selection behaviour of wild blackbirds at high prey densities. Heredity **32**, 401 – 409 (1974).

Harwood, R. H.: Predatory behaviour of *Argiope aurantia* (Lucas). Am. Midland Naturalist **91**, 130 – 139 (1974).

Hastings, J. W.: Light to hide by: Ventral luminescence to camouflage the silhouette. Science **173**, 1016 – 1017 (1971).

Hatch, J. J.: Predation and piracy by gulls at a ternery in Maine. Auk **87**, 244 – 254 (1970).

Hatch, J. J.: Piracy by laughing gulls (*Larus atricilla*): An example of the selfish group. MS 1972 or 1973.

Heatwole, H., Heatwole, A.: Motivational aspects of feeding behavior in toads. Copeia **4**, 692 – 698 (1968).

Hediger, H.: Putzerfische im Aquarium. Natur Museum **98**, 89 – 96 (1968).

Heiligenberg, W.: Ursachen für das Auftreten von Instinktbewegungen bei einem Fisch (*Pelmatochromis subocellatus kribensis* Boul., Cichlidae). Z. vergl. Physiol. **47**, 339 – 380 (1963).

Helbig, L.: Ethologische Beobachtungen an gefangengehaltenen *Egretta garzetta*, *Leucophoyx thula* und *Ardeola ibis* außerhalb der Brutzeit. Beitr. Vogelk. **13**, 397 – 454 (1968).

Henderson, B. A.: The specialized feeding behaviour of *Ambystoma gracile* in Marion Lake, Brit. Columbia. Can. Field Naturalist **87**, 151 – 154 (1973).

Hendrichs, H., Hendrichs, U.: Dikdik und Elefanten. München: Piper 1971.

Heppner, F.: Sensory mechanism and environmental clues used by the American robin in locating earthworms. Condor **67**, 246 – 256 (1965).

Herrlinger, E.: Die Wiedereinbürgerung des Uhus *Bubo bubo* in der Bundesrepublik Deutschland. Bonn. Zool. Monogr. **4**, 1 – 151 (1973).

Herting, G. E., Witt, A., Jr.: The role of physical fitness of forage fishes in relation to their vulnerability to predation by bowfin (*Amia calva*). Trans. Am. Fisheries Soc. **96**, 427 – 430 (1967).

Herzog, H. A., Burghardt, G. M.: Prey movement and predatory behaviour of juvenile western yellow-bellied racers, *Coluber constrictor mormon*. In press in Herpetologica.

Hespenheide, H. A.: Food preference and the extent of overlap in some insectivorous birds, with special reference to the Tyrannidae. Ibis **113**, 59 – 72 (1971).

Hespenheide, H. A.: A novel mimicry complex: beetles and flies, J. Entomol. **48**, 49 – 56 (1973).

Hess, E. H.: Imprinting. An effect of early experience, imprinting determines later social behaviour in animals. Science **130**, 133 – 141 (1959).

Hess, E. H.: Imprinting in birds. Science **146**, 1128 – 1139 (1964).

Heyder, R.: Gedächtnisleistung bei Vögeln. Beitr. Vogelkunde **16**, 192 – 194 (1970).

Himstedt, W., Schaller, F.: Versuche zu einer Analyse der Beutefang-Reaktionen von Urodelen auf optische Reize. Naturwissenschaften **53**, 619 (1966).

Hinde, R. A.: The behaviour of the great tit (*Parus major*) and some other related species. Behav. Suppl. **2,** 1 – 201 (1952).

Hinde, R. A.: Appetitive behaviour, consummatory act, and the hierarchical organization of behaviour, with special reference to the great tit (*Parus major*). Behav. **5,** 189 – 224 (1953).

Hinde, R. A.: Food and habitat selection in birds and lower vertebrates. 15th Intern. Congr. Zool., Sect. 10, Paper **18,** 1 – 2 (1958).

Hinde, R. A.: Unitary drives. Behav. **7,** 130 – 141 (1959).

Hinde, R. A.: Animal Behaviour. 2nd ed. New York: McGraw-Hill 1970.

Hindsbo, O.: Effects of *Polymorphus* (Acanthocephala) on colour and behaviour of *Gammarus lacustris*. Nature **238,** 333 (1972).

Hobson, E. S.: Predatory behaviour of some shore fishes in Gulf of California. U. S. Fish Wildlife Serv. Res. Rept. **73,** 1 – 92 (1968).

Hobson, E. S.: Cleaning symbiosis among California inshore fishes. Fisheries Bull. **69,** 491 – 523 (1971).

Hobson, E. S.: Activity of Hawaiian reef fishes during the evening and morning transitions between daylight and darkness. Fisheries Bull. **70,** 715 – 740 (1972).

Hogan, J. A.: An experimental study of conflict and fear: An analysis of behaviour of young chicks toward a mealworm. Part I. The behaviour of chicks which do not eat the mealworm. Behav. **25,** 45 – 97 (1965).

Hogan, J. A.: An experimental study of conflict and fear: An analysis of behaviour of young chicks toward a mealworm. Part II. The behaviour of chicks which eat the mealworm. Behav. **27,** 273 – 289 (1966).

Hogan, J. A.: The development of a hunger system in young chicks. Behav. **39,** 128 – 201 (1971).

Hölldobler, B.: Chemische Strategie beim Nahrungserwerb der Diebsameise (*Solenopsis fugax* Latr.) und der Pharaoameise (*Monomorium pharaonis* L.). Oecologia **11,** 371 – 380 (1973).

Holling, C. S.: The components of predation as revealed by a study of small-mammal predation of the European pine sawfly. Can. Entomologist **91,** 293 – 320 (1959).

Holling, C. S.: The functional response of predators to prey density and its role in mimicry and population regulation. Mem. Entomol. Soc. Canada **45,** 1 – 62 (1965).

Holling, C. S.: The functional response of invertebrate predators to prey density. Mem. Entomol. Soc. Canada **48,** 1 – 86 (1966).

Holmes, J. C., Bethel, W. M.: Modification of intermediate host behaviour by parasites. Behav. Asp. Parasite transmissions. Zool. J. Linnean Soc. **51,** 123 – 149 (1972).

Holst, E. v.: Quantitative Untersuchungen über Umstimmungsvorgänge im Zentralnervensystem. I. Der Einfluß des Appetits auf das Gleichgewichtsverhalten bei *Pterophyllum*. Z. vergl. Physiol. **31,** 134 – 148 (1948).

Hopkins, C. D., Wiley, R. H.: Food parasitism and competition in two terns. Auk **89,** 583 – 594 (1972).

Hoppenheit, M.: Untersuchungen über den Einfluß von Hunger und Sättigung auf das Beutefangverhalten der Larve von *Aeschna cyanea* Müll. (Odonata). Z. Wiss. Zool. **170,** 309 – 322 (1964 a).

Hoppenheit, M.: Beobachtungen zum Beutefangverhalten der Larve von *Aeschna cyanea* Müll. (Odonata). Zool. Anz. **172,** 216 – 232 (1964 b).

Hornocker, M. G.: An analysis of mountain lion predation upon mule deer and elk in the Idaho primitive area. Wildlife Monograph. **21,** 5 – 39 (1970).

Horstmann, E.: Form und Struktur von Starenschwärmen. Verhandl. Deut. Zool. Ges. 153 – 159 (1953).

Horstmann, K.: Untersuchungen zur Arbeitsteilung unter den Außendienstarbeiterinnen der Waldameise *Formica polyctena* Foerster. Z. Tierpsychol. **32,** 532 – 543 (1973 a).

Horstmann, K.: Untersuchungen zur Größenverteilung bei den Außendienstarbeiterinnen der Waldameise *Formica polyctena* Foerster (Hymenoptera, Formicidae). Waldhygiene **9**, 193 – 202 (1973 b).

Hughes, B. O., Wood-Gush, D. C. M.: A specific appetite for calcium in domestic chickens. Anim. Behav. **19**, 490 – 499 (1971).

Humphries, D. A., Driver, P. M.: Protean defence by prey animals. Oecologia **5**, 285 – 302 (1970).

Hundley, M. H.: Notes on methods of feeding and the use of tools in the Geospizinae. Auk **80**, 372 – 373 (1963).

Immelmann, K.: Periodische Vorgänge in der Fortpflanzung tierischer Organismen. Studium Gener. **20**, 15 – 33 (1967).

Immelmann, K.: Sexual and other long-term aspects of imprinting in birds and other species. Advan. Study Behav. **4**, 148 – 169 (1972).

Ingle, D.: Selective choice between double prey objects by frogs. Brain, Behav. Evol. **7**, 127 – 144 (1973).

Ingolfsson, A.: Behaviour of gulls robbing eiders. Bird Study **16**, 45 – 52 (1969).

Irving, L.: Arctic life of birds and mammals, including man. Zoophysiology and Ecology Vol. **2**, Berlin, Heidelberg, New York: Springer-Verlag 1972.

Isely, F. B.: Survival value of acridian protective coloration. Ecology **19**, 370 – 389 (1938).

Ivlev, V. S.: Experimental ecology of the feeding of fishes. New Haven: Yale Univ. Press 1961.

Jansson, E.: Anteckningar rörande häckande sparvuggla (*Glaucidium passerinum*). Vår Fågelv. **23**, 209 – 222 (1964).

Jenkins, D., Watson, A., Miller, G. R.: Population studies on red grouse, *Lagopus lagopus scoticus* (Lath.) in north-east Scotland. J. Anim. Ecol. **32**, 317 – 376 (1963).

Jenkins, D., Watson, A., Miller, G. R.: Predation and red grouse populations. J. Appl. Ecol. **1**, 183 – 195 (1964).

Jones, T. B., Kamil, A. C.: Tool-making and tool-using in the northern blue jay. Science **180**, 1076 – 1978 (1973).

Jouvet, M.: The states of sleep. Sci. American, **216**, 62 – 72 (1967).

Kabisch, K.: Zur Nestlingsnahrung von *Parus major* L. Arch. Forstwiss. **14**, 3 – 9 (1965).

Kabisch, K., Belter, H.: Das Verzehren von Amphibien und Reptilien durch Vögel. Staatl. Mus. Tierkd. Dresden **29**, 191 – 227 (1968).

Kahl, M. P., Peacock, L. J.: The bill-snap reflex: A feeding mechanism in the American wood stork. Nature **199**, 505 – 506 (1963).

Kalmijn, A. J.: The electric sense of sharks and rays. J. Exptl. Biol. **55**, 371 – 383 (1971).

Karli, P.: The Norway rat's killing response of the white mouse: An experimental analysis. Behav. **10**, 81 – 103 (1956).

Kaufman, D. W.: Shrike prey selection: Colour or conspicousness? Auk **90**, 204 – 206 (1973 a).

Kaufman, D. W.: Was oddity conspicuous in prey selection experiments? Nature **244**, 111 – 112 (1973 b).

Kaufman, D. W.: Differential predation on active and inactive prey by owls. Auk **91**, 172 – 173 (1974).

Kavanau, J. L., Ramos, J.: Twilight and onset and cessation of carnivore activity. J. Wildlife Management **2**, 653 – 657 (1972).

Kear, J.: Food selection in finches with special reference to interspecific differences. Proc. Zool. Soc. London **138**, 163 – 204 (1962).

Keenlyne, K. D.: Sexual differences in feeding habits of *Crotalus horridus horridus*. J. Herpetol. **6**, 234 – 237 (1972).

Kelsall, J. P.: The migratory barren ground caribou of Canada. In: Can. Wildlife Serv. Ottawa: Queen's Printer 1968.

Kettlewell, H. B. D.: Selection experiments on industrial melanism in the Lepidoptera. Heredity **9**, 323 – 342 (1955).

Kidder, G. W., Lilly D. M., Claff, C. L.: Growth studies on ciliates. IV. The influence of food on the structure of *Glaucoma vorax* Sp. Nov. Biol. Bull. **78,** 9 – 23 (1940).

Kleerekoper, H., Timms, A. M., Westlake, G. F., Davy, F. B., Malar, T., Anderson, V. M.: An analysis of locomotor behaviour of goldfish (*Carassius auratus*). Anim. Behav. **18,** 317 – 330 (1970).

Klingauf, F., Sengonca, C.: Koloniebildung von Röhrenblattläusen (Aphididae) unter Feindeinwirkung. Entomophaga **15,** 359 – 377 (1970).

Klopfer, P.: Behavioural aspects of ecology. Englewood Cliffs, New Jersey: Prentice-Hall. Inc. 1962.

Klopfer, P.: Ökologie und Verhalten. Stuttgart: Fischer 1968.

Kniprath, E.: Nahrung und Nahrungserwerb des Eisvogels. Vogelwelt **90,** 81 – 97 (1969).

Koehler, O.: „Zählende" Vögel und vergleichende Verhaltensforschung. Acta 11th Congr. Intern. Ornithol. 1954, 588 – 598 (1955).

Koenig, L.: Beiträge zu einem Aktionssystem des Bienenfressers (*Merops apiaster* L.). Z. Tierpsychol. **8,** 169 – 210 (1951).

Kortlandt, A.: Discussion contribution to: Hall, K. R. L., Too-using performances as indicators of behavioural adaptability. Current. Anthropol. **7,** 215 – 216 (1966).

Kortlandt, A.: New Perspectives on Ape and Human Evolution. Amsterdam: Stichting voor Psychobiologie 1972.

Kortlandt, A., Kooij, M.: Protohominid behaviour in primates (preliminary communication). Symp. Zool. Soc., London **10,** 61 – 88 (1963).

Kramer, G.: Beobachtungen über das Verhalten der Aaskrähe *Corvus corone* zu Freund und Feind. Festschrift Oskar Heinroth, J. Orn., 105 – 131 (1941).

Krames, L., Milgram, N. W., Christie, D. P.: Predatory agression: Differential suppression of killing and feeding. Behav. Biol. **9,** 641 – 647 (1973).

Krebs, J. R.: Behavioural aspects of predation. In: Perspectives in Ethology, 73 – 111. P. P. G. Bateson and P. H. Klopfer (eds.), New York: Plenum Press 1973 a.

Krebs, J. R.: Social learning and the significance of mixed-species flocks of chickadees (*Parus* spp.). Can. J. Zool. **51,** 1275 – 1288 (1973 b).

Krebs, J. R., MacRoberts, M. H., Cullen, J. M.: Flocking and feeding in the great tit *Parus major* – an experimental study. Ibis **114,** 507 – 530 (1972).

Krieckhaus, E. E., Wolf, G.: Acquisition of sodium by rats: Interaction of innate mechanisms and latent learning. J. Comp. Physiol. Psychol. **65,** 197 – 201 (1968).

Kruuk, H.: Predators and anti-predator behaviour of the blackheaded gull (*Larus ridibundus* L.). Behav. **11,** 1 – 129 (1964).

Kruuk, H.: Surplus killing by carnivores. J. Zool. London **166,** 233 – 244 (1972 a).

Kruuk, H.: The Spotted Hyena. Chicago, London: Univ. Chic. Press 1972 b.

Kühme, W.: Freilandstudien zur Soziologie des Hyänenhundes (*Lycaon pictus lupinus* Thomas 1902). Z. Tierpsychol. **22,** 495 – 541 (1965).

Kushlan, J. A.: Bill-vibrating. A prey-attracting behaviour of the snowy egret, *Leucophoyx thula*. Am. Midland Naturalist **89,** 509 – 512 (1973).

Kuyt, E.: Food habits of wolves on barren-ground caribou range. Can. Wildlife Serv. Rept. Ser. **21,** 1 – 35 (1972).

Lack, D.: Competition for food by birds of prey. J. Animal Ecol. **15,** 123 – 129 (1946).

Lack, D.: The natural regulation of animal numbers. Oxford: Oxford Univ. Press 1954.

Lack, D.: Population studies of birds. Oxford: Clarendon Press, Oxford Univ. Press 1966.

Lack, D.: Ecological isolation in birds. Oxford, Edinburgh: Blackwell Sci. Publ. 1971.

Laing, J.: Host-finding by insect parasites. I. Observations on the finding of hosts by *Alysia manducator, Mormoniella vitripennis* and *Trichogramma evanescens*. *J.* Anim. Ecol. **6,** 298 – 317 (1937).

Laing, J.: Host-finding by insect parasites. II. The chance of *Trichogramma evanescens* finding its hosts. J. Exptl. Biol. **15,** 281 – 302 (1938).

223

Lambert, G.: Predation efficiency of the osprey. Can. Field Naturalist **57**, 87 – 88 (1943).

Landry, St. O.: The rodentia as omnivores. Quart. Rev. Biol. **45**, 351 – 372 (1970).

Landsborough Thomson, A. (ed.): A New Dictionary of Birds. London, Edinburgh: Nelson 1964.

Lange, G.: Über Nahrung, Nahrungsaufnahme und Verdauungstrakt mitteleuropäischer Limikolen. Beitr. Vogelkd. **13**, 225 – 334 (1968).

Lawick-Goodall, H. van, Lawick-Goodall, J. van: Innocent Killers, London: Collins, 1970.

Lawick-Goodall, J. van: Behaviour of free-living chimpanzees of the Gombe Stream area. Anim. Behav. Monogr. **3** (1968).

Lawick-Goodall, J. van, Lawick-Goodall, H. van: Use of tools by the Egyptian vulture, *Neophron percnopterus*. Nature **212**, 1468 – 1469 (1966).

Lazarus, J.: Natural selection and the functions of flocking in birds: A reply to Murton. Ibis **114**, 556 – 558 (1972).

Lehr, E.: Experimentelle Untersuchungen an Affen und Halbaffen über Generalisation von Insekten- und Blütenabbildungen. Z. Tierpsychol. **24**, 208 – 244 (1967).

Lehr, E.: Visuelle Orientierung bei Primaten und damit verbundene Lern- und Abstraktionsleistungen. Fortschr. Zool. **21**, 334 – 348 (1973).

Leyhausen, P.: Über die Funktion der relativen Stimmungshierarchie. Dargestellt am Beispiel der phylogenetischen und ontogenetischen Entwicklung des Beutefangs von Raubtieren. Z. Tierpsychol. **22**, 412 – 494 (1965 a).

Leyhausen, P.: The communal organization of solitary mammals. Symp. Zool. Soc. London. **14**, 249 – 263 (1965 b).

Leyhausen, P.: Verhaltensstudien an Katzen. 3rd ed. 1973. Berlin, Hamburg: Paul Parey.

Ligon, J. D.: The biology of the elf owl, *Micrathene whitneyi*. Misc. Publ. Museum. Zool., Univ. Mich. **136**, 1 – 70 (1968).

Lind, H.: Parental feeding in the oystercatcher (*Haematopus o. ostralegus* L.). Dansk Orn. Foren. Tidsskr. **59**, 1 – 31 (1965).

Lindstedt, K. J.: Chemical control of feeding behaviour. Comp. Biochem. Physiol. **39 A**, 553 – 581 (1971).

Lindstrom, R., Nilsson, N. A.: On the competition between whitefish species. In: The Exploitation of Natural Animal Populations. Le Cren, M. W. Holdgate (eds.), Symp. Brit. Ecol. Soc. **2**, 326 – 340 (1962).

Lindroth, C. H.: Disappearance as a protective factor: A supposed case of Bates'ian mimicry among beetles (Coleoptera: Carabidae, Chrysomelidae). Entomol. Scand. **2**, 41 – 48 (1971).

Lineaweaver, T. H., Backus, R. H.: The natural history of sharks. London: Deutsch Ltd. 1970.

Lloyd, M., Dybas, H. S.: The periodical cicada problem. I. Population ecology. II. Evolution. Evolution **20**, 133 – 149, 466 – 505 (1966).

Löhrl, H.: Zum Verhalten der Rabenkrähe (*Corvus c. corone*) gegenüber dem Habicht. Z. Tierpsychol. **7**, 130 – 133 (1950 a).

Löhrl, H.: Zur Biologie des Kuckucks. Orn. Ber. **3**, 120 – 125 (1950 b).

Löhrl, H.: Welche Meisenarten verstecken Futter? Vogelwelt **76**, 210 – 212 (1955).

Löhrl, H.: Weißstorch „erschnäbelt" Beute im Wasser. Vogelwarte **19**, 52 – 53 (1957).

Löhrl, H.: *Martes foina* (Mustelidae) Sichern. Encycl. Cinemat. 1 – 6 (1972).

Löhrl, H.: Die Tannenmeise. Wittenberg Lutherstadt: Ziemsen 1974.

Łomnicki, A., Slobodkin, L. B.: Floating in *Hydra littoralis*. Ecology **47**, 881 – 889 (1966).

Loop, M. S.: Twilight movement patterns of some florida snakes and potential predators. ASB Bull. **19**, 81 (1972).

Loop, M. S., Scoville, S. A.: Response of newborn *Eumeces inexpectatus* to prey-object extracts. Herpetologica **28**, 254 – 256 (1972).

Lüling, K. H.: *Colisa lalia* (Anabantidae) Beutespucken. Encycl. Cinemat. E 1674. Göttingen: Inst. Wiss. Film 1972.

224

Lüling, K. H.: *Arapaima gigas* (Osteoglossidae) Beutefang. Encycl. Cinemat. E 1692. Göttingen: Inst. Wiss. Film 1973.

MacArthur, R. H.: Geographical Ecology. New York: Harper and Row 1972.

Macura, A.: Delayed reactions in the tawny owl (*Strix aluco* L.). Folia Biol. **7**, 330 – 337 (1959).

Magnus, D. B. E.: Zum Problem der Partnerschaften mit Diademseeigeln. Verhandl. Deut. Zool. Ges. München 1963.

Mahmoud, J. Y., Lavenda, N.: Establishment and eradication of food preferences in red-eared turtles. Copeia **2**, 298 – 300 (1969).

Maldonado, H.: The control of attack by *Octopus*. Z. vergl. Physiol. **47**, 656 – 674 (1964).

Maldonado, H., Benko, M., Isern, M.: Study of the role of the binocular vision in mantids to estimate long distances, using the deimatic reactions as experimental situation. Z. vergl. Physiol. **68**, 72 – 83 (1970).

Maldonado, H., Rodriguez, E., Balderrama, N.: How mantids gain insight into the new maximum catching distance after each ecdysis. J. Insect Physiol. **20**, 591 – 603 (1974).

Maly, E. J.: The influence of predation on the adult sex ratios of two copepod species. Limnol. Oceanog. **15**, 566 – 573 (1970).

Manteifel, B. P., Radakov, D. V.: Study of the adaptive value of schooling behaviour in fish. Usp. Sovrem. Biol. **50**, No. 3 (1960).

Markl, H.: Aggression und Beuteverhalten bei Piranhas (Serrasalminae, Characidae). Z. Tierpsychol. **30**, 190 – 216 (1972).

Marler, P.: Specific distinctiveness in the communication signals of birds. Behav. **11**, 13 – 39 (1957).

Martin, J. B., Witherspoon, N. B., Keenleyside, M. H. A.: Analysis of feeding behaviour in the newt *Notophthalmus viridescens*. Can. J. Zool. **52**, 277 – 281 (1974).

Martinez, D. R., Klinghammer, E.: The behaviour of the whale *Orcinus orca*: A review of the literature. Z. Tierpsychol. **27**, 828 – 839 (1970).

Mason, L. G.: Prey selection by a non-specific predator. Evolution **19**, 259–260 (1965).

Mayer, K., Quednau, W.: Der Einfluß des Wirtes auf das Verhalten parasitärer Insekten. 15th. Intern. Zool. Congr., Sect. VIII, No. 22, London 1958.

Mayer, K., Quednau, W.: Verhaltensänderungen bei Eiparasiten der Gattung *Trichogramma* unter dem Einfluß des Wirtes. Z. Parasitenkd. **19**, 35 – 41 (1959).

Maynard-Smith, J.: The causes of polymorphism. Symp. Zool. Soc. London **26**, 371 – 383 (1970).

Mayr, E.: Animal species and evolution. Cambridge, Mass.: Harvard Univ. Press 1963.

McKinney, F.: An analysis of the displays of the European eider *Somateria mollissima mollissima* (Linnaeus) and the Pacific eider *Somateria mollissima v. nigra* Bonaparte. Behav. Suppl. **7**, 1 – 124 (1961).

McNicholl, M. K.: Habituation of aggressive responses to avian predators by terns. Auk **90**, 902 – 904 (1973).

Mead, R. A.: Cooper hawk attacks pigeon by stooping. Condor **65**, 167 (1963).

Mebs, T.: Wanderfalke versteckt Beute. Vogelwelt **77**, 12 – 15 (1956).

Mech, L. D.: The Wolf: The Ecology and Behaviour of an Endangered Species. Garden City, N. Y.: Natural History Press 1970.

Mech, L. D., Frenzel, L. D., Jr. (eds.): Ecological studies of the timber wolf in Northeastern Minnesota. U. S. Dept. Agr. Forest Ser. Res. Paper NC **52**, 1 – 62 (1971).

Meesters, A.: Über die Organisation des Gesichtsfeldes der Fische. Z. Tierpsychol. **4**, 84 – 149 (1941).

Meinertzhagen, R.: Pirates and Predators. Edinburgh and London: Oliver and Boyd 1959.

Mendelssohn, H., Golani, I., Marder, U.: Agricultural development and the distribution of venomous snakes and snake bite in Israel. In: Toxins of Animal and Plant Origin. A. de Vries, E. Kochva (eds.), London: Gordon and Breach 1971.

Menzel, E. E.: Chimpanzee spatial memory organization. Science **182**, 943 – 945 (1973).

Messenger, J. B.: The visual attack of the cuttlefish, *Sepia officinalis*. Anim. Behav. **16**, 342 – 357 (1968).

225

Metzger, L. H.: An experimental comparison of screech owl predation on resident and transient whitefooted mice (*Peromyscus leucopus*). J. Mammal. **48**, 387 – 391 (1967).
Meyer-Holzapfel, M.: Angeborene und erworbene Beutemerkmale beim Waldkauz. J. Orn. **94**, 376 (1953).
Meyerriecks, A. J.: Foot-stirring behaviour in herons. Wils. Bull. **71**, 153 – 158 (1959).
Mikkola, H.: On the activity and food of the pygmy owl *Glaucidium passerinum* during breeding. Ornis Fenn. **47**, 10 – 14 (1970).
Milinski, M., Curio, E.: Untersuchungen zur Selektion durch Räuber gegen Vereinzelung der Beute. Z. Tierpsychol. **37**, 400 – 402 (1975).
Miller, N. E.: Experiments on motivation. Studies combining psychological, physiological and pharmacological techniques. Science **126**, 1271 – 1278 (1957).
Millikan, G. C., Bowman, R. I.: Observations on Galapagos tool-using finches in captivity. Living Bird **6**, 23 – 41 (1967).
Mittelstaedt, H.: Prey capture in mantids. In: Recent advances in invertebrate physiology. B. T. Scheer (ed.), Eugene, Ore.: Univ. Ore, Publ. 51 – 71, 1957.
Molitor, A.: Das Verhalten der Raubwespen. Z. Tierpsychol. **3**, 60 – 74 (1939).
Molle, F.: Anolis fressen Gehäuseschnecken. Deut. Aquar.-Terrarien-Z. **7**, 307 – 308 (1957).
Moment, G.: Reflexive selection: A possible answer to an old puzzle. Science **136**, 262 – 263 (1962).
Mook, J. H., Mook, L. J., Heikens, H. S.: Further evidence for the role of "searching images" in the hunting behaviour of titmice. Arch. Néerl. Zool. **13**, 448 – 465 (1960).
Moran, Sh., Fishelson, L.: Predation of a sand-dwelling mysid crustacean *Gastrosaccus sanctus* by plover birds (Charadriidae). Marine Biol. **9**, 63 – 64 (1971).
Morrell, G. M., Turner, J. R. G.: Experiments on mimicry: The response of wild birds to artificial prey. Behav. **36**, 116 – 130 (1970).
Morris, R. F.: The effect of predator age and prey defense on the functional response of *Podisus maculiventris* Say to the density of *Hyphantria cunea* Drury. Can. Entomologist **95**, 1009 – 1020 (1963).
Morris, D. D., Loop, M. S.: Stimulus control of prey attack in naive rat snakes: A species duplication. Psychon. Sci. **15**, 141 – 142 (1969).
Morrison, G. D.: Notes on a lizard, *Tarentola* sp. (Gekkonidae) found in Aberdeen. Am. Mag. Nat. Hist. 10th Ser. **20**, 315 – 318 (1937).
Morrison, G. R.: Alterations in palatability of nutrients for the rat as result of prior tasting. J. Comp. Physiol. Psychol. **86**, 56 – 61 (1974).
Mortensen, T.: Observations on protective adaptations and habits, mainly in marine animals. Vidensk. Medd. Dansk Naturhist. Foren. **69**, 57 – 96 (1917).
Morton, E. S.: On the evolutionary advantages and disadvantages of fruit eating in tropical birds. Am. Naturalist **107**, 8 – 22 (1973).
Moynihan, M.: The organization and probable evolution of some mixed species flocks of neotropical birds. Smithson. Inst. Misc. Collections **143**, 1 – 140 (1962).
Moynihan, M.: Social mimicry; character convergence versus character displacement. Evolution **22**, 315 – 331 (1968).
Mueller, H. C.: Prey selection: Oddity or conspicuousness? Nature **217**, 92 (1968).
Mueller, H. C.: Oddity and specific searching image more important than conspicuousness in prey selection. Nature 233, 345 – 346 (1971).
Mueller, H. C.: Further evidence for the selection of odd prey by hawks. Am. Zoologist **12**, 656 (1972).
Mueller, H. C.: The relationship of hunger to predatory behaviour in hawks (*Falco sparverius* and *Buteo platypterus*). Anim. Behav. **21**, 513 – 520 (1973).
Mueller, H. C.: Food caching behaviour in the American kestrel (*Falco sparverius*). Z. Tierpsychol. **34**, 105 – 114 (1974).
Mueller, H. C.: Hawks select odd prey. Science **188**, 953 – 954 (1975).
Münster, W.: Der Neuntöter oder Rotrückenwürger. In: Neue Brehm-Bücherei **218**, Wittenberg Lutherstadt: A. Ziemsen 1958.

Murdoch, W. W.: Switching in general predators: Experiments on predator specificity and stability of prey populations. Ecol. Mongr. **39**, 335 – 354 (1969).

Murdoch, W. W.: The developmental response of predators to changes in prey density. Ecology **52**, 132 – 137 (1971).

Murdoch, W. W., Marks, J. R.: Predation by coccinellid beetles: Experiments on switching. Ecology **54**, 160 – 167 (1973).

Murdock, G. R., Murdock, B. S.: Tentacle length and prey capture as a function of starvation time in hydra. Am. Zoologist **12**, 719 (1972).

Murie, A.: Following fox trails. Misc. Publ., Univ. Mich. Mus. Zool. **32**, 1 – 45 (1936).

Murie, A.: The wolves of Mount McKinley. Fauna of the National Parks of the United States, Fauna Ser. No. **5**, 1944.

Murton, R. K.: The significance of a specific search image in the feeding behaviour of the wood-pigeon. Behav. **40**, 10 – 42 (1971).

Murton, R. K., Isaacson, A. J., Westwood, N. J.: The significance of gregarious feeding behaviour and adrenal stress in a population of wood-pigeons *Columba palumbus*. J. Zool. **165**, 53 – 84 (1971).

Myer, J. S., Kowell, A. P.: Eating patterns and body weight change of snakes when eating and when food deprived. Physiol. Behav. **6**, 71 – 74 (1971).

Myrberget, S., Groven, B., Myhre, R.: Tracking wolverines, *Gulo gulo,* in the Jotunheim mountains, South Norway, 1965 – 68. Fauna **22**, 237 – 252 (1969).

Mysterud, I.: Behaviour of the brown bear (*Ursus arctos*) at moose kills. Norweg. J. Zool. **21**, 267 – 272 (1973).

Napier, J. R.: Early man and his environment. Discovery, March, 12 – 18 (1963).

Naulleau, G.: La biologie et le comportement prédateur de *Vipera aspis* au laboratoire et dans la nature. Bull. Biol. Fr. Belg. **99**, 397 – 524 (1966).

Neill, S. R.: A study of anti-predator adaptation in fish with special reference to silvery camouflage and shoaling. Ph. D. thesis, Oxford (1970).

Neill, S. R., Cullen, J. M.: Experiments on whether schooling by their prey affects the hunting behaviour of cephalopod and fish predators. J. Zool. London **172**, 549 – 569 (1974).

Neill, W. T.: The Last of the Ruling Reptiles. Alligators, Crocodiles, and their Kin. New York and London: Columbia Univ. Press 1971.

Neish, I. C.: A comparative analysis of the feeding behaviour of two salamander populations in Marion Lake, B. C. Ph. D. thesis, Univ. Brit. Columbia (1970).

Neisser, U.: Visual search. Sci. Am. **210**, 94 – 102 (1964).

Neisser, U.: Cognitive Psychology. New York: Appleton-Century-Crofts 1966.

Nelson, D. R., Gruber, S. H.: Sharks: Attraction by low-frequency sounds. Science **142**, 975 – 977 (1963).

Nilsson, N. A.: Seasonal fluctuations in the food segregation of trout, char, and whitefish in 14 North-Swedish Lakes. Rept. Inst. Freshwater Res. Drottningholm **41**, 185 – 205 (1960).

Nisbet, I. C. T.: Courtship-feeding, egg-size and breeding success in common terns. Nature **241**, 141 – 142 (1973).

Njine, T.: La transformation microstome-macrostome et macrostome-microstome chez *Tetrahymena paravorax* Corliss 1957. Ann. Fac. Sci. Cameroun **10**, 69 – 84 (1972).

Norberg, R. A.: Hunting technique of Tengmalm's owl, *Aegolius funereus*. Ornis Scand. **1**, 51 – 64 (1970).

Norton-Griffiths, M.: The feeding behaviour of the oystercatcher (*Haematopus ostralegus*). Ph. D. thesis, Oxford (1968).

Norton-Griffiths, M.: The organization, control and development of parental feeding in the oystercatcher (*Haematopus ostralegus*). Behav. **34**, 55 – 114 (1969).

Noton, D., Stark, L.: Eye movements and visual perception. Sci. Am. **224**, 34 – 43 (1971).

Nursall, J. R.: Some behavioural interactions of spottail shiners (*Notropis hudsonius*), yellow perch (*Perca flavescens*), and northern Pike (*Esox lucius*). J. Fisheries Res. Board Can. **30**, 1161 – 1178 (1973).

Nyberg, D. W.: Prey-capture in the largemouth bass. Am. Midland Naturalist **86**, 128 – 144 (1971).
O'Donald, P., Pilecki, C.: Polymorphic mimicry and natural selection. Evolution **24**, 395 – 401 (1970).
Ogle, T. F.: Predator-prey relationships between coyotes and white-tailed deer. Northwest. Sci. **45**, 213 – 218 (1971).
Olla, B. L., Samet, C.: Fish-to-fish attraction and the facilitation of feeding behaviour as mediated by visual stimuli in striped mullet, *Mugil cephalus.* J. Fish. Res. Board Can. **31**, 1621 – 1630 (1974).
Osche, G.: Ökologie des Parasitismus und der Symbiose (einschl. Phoresie, Epökie und Kommensalismus). Fortschr. Zool. **15**, 125 – 164 (1963).
Owen, D. F.: Polymorphism and population density in the African land snail *Limicolaria martensiana.* Science **140**, 666 – 667 (1963).
Owen, D. F.: Density effects in polymorphic land snails. Heredity **20**, 312 – 315 (1965).
Paine, R. T.: Food recognition and predation on opisthobranchs by *Navanax inermis* (Gastropoda: Opisthobranchia). Veliger **6**, 1 – 9 (1963).
Palm, P.-O.: Rödrävens näringsekologi. Zool. Rev. **32**, 43 – 46 (1970).
Paloheimo, J. E.: On a theory of search. Biometrica **58**, 61 – 75 (1971).
Paris, O. H.: Some quantitative aspects of predation by muricid snails on mussels in Washington Sound. Veliger **2**, 41 – 47 (1960).
Parsons, P. A.: Behavioural and ecological genetics. Oxford: Clarendon Press 1973.
Paul, L.: Predatory attack by rats: Its relationship to feeding and type of prey. J. Comp. Physiol. Psychol. **78**, 69 – 76 (1972).
Paul, L., Posner, I.: Predation and feeding: Comparisons of feeding behaviour of killer and nonkiller rats. J. Comp. Physiol. Psychol. **84**, 258 – 264 (1973).
Payne, R. S.: How the barn owl locates prey by hearing. Living Bird **1**, 151 – 159 (1962).
Peeters, H. J.: Einiges über den Waldfalken *Micrastur semitorquatus.* J. Orn. **104**, 357 – 364 (1963).
Peregoy, P. L., Zimmermann, R. R., Strobel, D. A.: Protein preference in protein-malnourished monkeys. Percept. Motor Skills **35**, 495 – 503 (1972).
Pielowski, Z.: Studies on the relationship: predator (goshawk) prey (pigeon). Bull. Acad. Polon. Sci. Ser. Sci. Biol. **7**, 401 – 403 (1959).
Pielowski, Z.: Über den Unifikationseinfluß der selektiven Nahrungswahl des Habichts (*Accipiter gentilis* L.) auf Haustauben. Ekologia Polska – Ser. A, Tom **IX**, 183 – 192 (1961).
Pilecki, C., O'Donald, P.: The effects of predation on artificial mimetic polymorphisms with perfect and imperfect mimics at varying frequencies. Evolution **25**, 365 – 370 (1971).
Pilleri, G., Knuckey, J.: Behaviour patterns of some Delphinidae observed in the Western Mediterranean. Z. Tierpsychol. **26**, 48 – 72 (1969).
Pimlott, D. H., Shannon, J. A., Kolenosky, G. B.: The ecology of the timber wolf. Res. Branch Res. Rept. (Wildlife) No. **87** (1969).
Pitelka, F. A., Tomich, P. Q., Treichel, G. W.: Ecological relations of jaegers and owls as lemming predators near Barrow, Alaska. Ecol. Monogr. **25**, 85 – 117 (1955).
Popham, E. J.: The variation in the colour of certain species of *Arctocorisa* (Hemiptera, Corixidae) and its significance. Proc. Zool. Soc. London **111**, 135 – 172 (1942).
Potts, G. W.: The schooling ethology of *Lutianus monostigma* (Pisces) in the shallow reef environment of Aldabra. J. Zool. London **161**, 223 – 235 (1970).
Potts, G. W.: The ethology of *Labroides dimidiatus* (Cuv. & Val.) (Labridae, Pisces) on Aldabra, Anim. Behav. **21**, 250 – 291 (1973 a).
Potts, G. W.: Cleaning symbiosis among British fish with special reference to *Crenilabrus melops* (Labridae). J. Marine Biol. Assoc. U. K. **53**, 1 – 10 (1973 b).
Pramer, D.: Nematode-trapping fungi. Science **144**, 382 – 388 (1964).

Precht, H., Freytag, G.: Über Ermüdung und Hemmung angeborener Verhaltensweisen bei Springspinnen (Salticidae). Zugleich ein Beitrag zum Triebproblem. Behav. **13**, 143 – 211 (1958).

Price, P.: Trail odors: Recognition by insects parasitic on cocoons. Science **170**, 546 – 547 (1970).

Pritchard, R. C. H.: Sea turtles of the Guianas. Bull. Florida State Mus., Biol. Sci. **13**, 85 – 140 (1969).

Rabinowitch, V. E.: The role of early experience in the development and retention of food habits in some birds. Ph. D. thesis, Univ. Wisconsin 1965.

Rabinowitch, V. E.: The role of experience in the development of food preferences in gull chicks (*Larus argentatus, L. delawarensis*). Anim. Behav. **16**, 425 – 428 (1968).

Radakov, D. V.: Schooling in the Ecology of Fish. Transl. from Russsian by H. Mills. New York: Wiley 1973.

Radu, D.: Die Aufzucht von Großtrappen (*Otis tarda*) im Zoologischen Garten Bukarest. Freunde des Kölner Zoo **12**, 59 – 64 (1969).

Räber, H.: Das Verhalten gefangener Waldohreulen und Waldkäuze zur Beute. Behav. **2**, 1 – 95 (1949).

Rand, A. L.: Foot-stirring as a feeding habit of wood ibis and other birds. Am. Midland Naturalist **55**, 96 – 100 (1956).

Rapport, D. J.: An optimization model of food selection. Am. Naturalist **105**, 575 – 587 (1971).

Rasa, O. A. E.: Prey capture, feeding techniques, and their ontogeny in the African dwarf mongoose, *Helogale undulata rufula*. Z. Tierpsychol. **32**, 449 – 488 (1973).

Rasmussen, K.: The Netsilik Eskimos: Social life and Spiritual culture. Rep. Fifth Thule Expedition, vol. 8, 1931 (Kopenhagen).

Reese, E. S.: Ethology and marine zoology. Oceanogr. Marine Biol. Ann. Rev., 455 – 488 (1964).

Remmert, H.: Tageszeitliche Verzahnung der Aktivitäten verschiedener Organismen. Oecologia **3**, 214 – 226 (1969).

Richter, C. P.: The self-selection of diets. In: Essays in Biology. Berkeley: Univ. Calif. Press, 501 – 506, 1943 a.

Richter, C. P.: Total self-regulatory functions in animals and human beings. Harvey Lectures, Ser. **38**, 63 – 103 (1943 b).

Riehm, H.: Ökologie und Verhalten der Schwanzmeise (*Aegithalos caudatus* L.). Zool. Jb. Syst. **97**, 338 – 400 (1970).

Rilling, S., Mittelstaedt, H., Roeder, K. O.: Prey recognition in the praying mantis. Behav. **14**, 164 – 184 (1959).

Ritte, U.: Floating and sexuality in laboratory populations of *Hydra littoralis*. Ph. D. thesis, Univ. Mich., Ann Arbor 1969.

Roberts, T. R.: Ecology of fishes in the Amazon and Congo basins. Bull. Museum Comp. Zool. **143**, 117 – 147 (1972).

Robinson, M. H.: Anti-predator adaptations in stick and leaf mimicking insects. Ph. D. thesis, Univ. Oxford, Radcliffe Library No. D. 3808, 1966.

Robinson, M. H.: The defensive behaviour of the stick insect *Oncotophasma martini* (Griffini) (Orthoptera; Phasmatidae). Proc. Roy. Entomol. Soc. London **43**, 183 – 187 (1968).

Robinson, M. H.: Defenses against visually hunting predators. In: Evolutionary Biology 3. T. Dobzhansky, M. K. Hecht, W. C. Steere (eds.), New York: Appleton-Century-Crofts, Educational Division, Meredith Corporation 225 – 259, 1969 a.

Robinson, M. H.: Predatory behaviour of *Argiope argentata* (Fabricius). Am. Zoologist **9**, 161 – 173 (1969 b).

Robinson, M. H.: Insect anti-predator adaptations and the behaviour of predatory primates. Congr. Latin. Zool. **II**, 811 – 836 (1970).

Rodgers, W. L.: Specificity of specific hungers. J. Comp. Physiol. Psychol. **64**, 49 – 58 (1967).

Roeder, K. D.: Neurale Grundlagen des Verhaltens. Bern und Stuttgart: Huber 1968.

Roeder, K. D., Treat, A. E.: The detection and evasion of bats by moths. Am. Scientist **49**, 135 – 148 (1961).

Root, R. B.: The niche exploitation pattern of the blue-gray gnatcatcher. Ecol. Monogr. **37**, 317 – 350 (1967).

Rosenthal, H.: Verdauungsgeschwindigkeit, Nahrungswahl und Nahrungsbedarf bei den Larven des Herings *Clupea harengus* L. Ber. Deut. Wiss. Komm. Meeresforsch. **20**, 60 – 69 (1969 a).

Rosenthal, H.: Untersuchungen über das Beutefangverhalten bei Larven des Herings *Clupea harengus*. Marine Biol. **3**, 208 – 221 (1969 b).

Rosenthal, H., Hempel, G.: Experimental studies in feeding and food requirements of herring larvae (*Clupea harengus* L.). In: Marine Food Chains. J. H. Steele (ed.), Edinburgh: Oliver and Boyd, 344 – 364, 1970.

Roth, V. D.: Unusual predatory activities of Mexican jays and brown-headed cowbirds under conditions of deep snow in southeastern Arizona. Condor **73**, 113 (1971).

Rothschild, M.: Is the buff ermine (*Spilosoma lutea* Huf) a mimic of the white ermine (*Spilosoma lubricipeda* L.)? In: General Entomology, Proc. Roy. Entomol. Soc. London **38**, 159 – 164 (1963).

Rothschild, M.: Speculations about mimicry with Henry Ford. In: Ecological Genetics and Evolution. R. Creed (ed.), 202 – 223 (1971).

Rothschild, M., Clay, T.: Fleas, Flukes, and Cuckoos. London: Collins 1957.

Roughgarden, J.: The fundamental and realized niche of a solitary population. Am. Naturalist **108**, 232 – 235 (1974).

Royama, T.: Factors governing the hunting behaviour and selection of food by the great tit (*Parus major* L.). J. Animal. Ecol. **39**, 619 – 668 (1970 a).

Royama, T.: Evolutionary significance of predators' response to local differences in prey density: A theoretical study. Proc. Advan. Study Inst. Dynamics Numbers Popul. (Oosterbeek, 1970 b) 344 – 357.

Rozin, P., Kalat, J. W.: Specific hungers and poison avoidance as adaptive specializations of learning. Psychol. Rev. **78**, 459 – 486 (1971).

Rubinoff, I., Kropach, C.: Differential reactions of atlantic and pacific predators to sea snakes. Nature **228**, 1288 – 1290 (1970).

Rudebeck, G.: The choice of prey and modes of hunting of predatory birds with special reference to their selective effect. Oikos **2**, 66 – 88 (1950).

Rudebeck, G.: The choice of prey and modes of hunting of predatory birds with special reference to their selective effect. The merlin (*Falco columbarius* L.). Oikos **3**, 200 – 231 (1951).

Rüppell, G., Gösswein, E.: Die Schwärme von *Leucaspius delineatus* (Cyprinidae, Teleostei) bei Gefahr im Hellen und im Dunkeln. Z. vergl. Physiol. **76**, 333 – 340 (1972).

Ruiter, L. de: Some experiments on the camouflage of stick caterpillars. Behav. **4**, 222 – 232 (1952).

Ruiter, L. de: Countershading in caterpillars. An analysis of its adaptive significance. Arch. Néerl. Zool. **11**, 1 – 57 (1955).

Ruiter, L. de: The physiology of vertebrate feeding behaviour towards a synthesis of the ethological and physiological approaches to problems of behaviour. Z. Tierpsychol. **20**, 498 – 516 (1963).

Ruiter, L. de: Feeding behaviour of vertebrates in the natural environment. In: Handbook of Physiology. Alimentary Canal **7**. Ed. by C. F. Code, Am. Physiol. Soc. 1967.

Russell, E.: An investigation of the palatability of some marine invertebrates to four species of fish. Pacific Sci. **20**, 452 – 460 (1966).

Salt, G. W.: Feeding activity by *Amoeba proteus*. Exptl. Cell. Res. **24**, 618 – 620 (1961).

Salt, G. W.: Predation in an experimental protozoan population (*Woodruffia-Paramecium*). Ecol. Monogr. **37**, 113 – 114 (1967).

Salt, G. W., Willard, D. E.: The hunting behaviour and success of Forster's tern. Ecology **52**, 989 – 998 (1971).

Sandness, J. N., McMurtry, J. A.: Prey consumption of *Amblyseius largoensis* in relation to hunger. Can. Entomologist **104**, 461 – 470 (1972).

Sauer, E. G. F.: Die Entwicklung der Lautäußerungen vom Ei ab schalldicht gehaltener Dorngrasmücken (*Sylvia c. communis* Latham) im Vergleich mit später isolierten und mit wildlebenden Artgenossen. Z. Tierpsychol. **11**, 10 – 93 (1954).

Schaller, G. B.: The Deer and the Tiger. Chicago, London: Univ. Chicago Press 1967.

Schaller, G. B.: This gentle and elegant cat. Nat. Hist. June/July **79**, 30 – 39 (1970).

Schaller, G. B.: The Serengeti Lion. Univ. Chicago Press, 1972.

Schaller, G. B., Lowther, G. R.: The relevance of carnivore behaviour to the study of early hominids. Southwestern J. Anthropol. **25**, 307 – 341 (1969).

Scherzinger, W.: Zum Aktionssystem des Sperlingskauzes (*Glaucidium passerinum* L.). Zoologica **118**, 1 – 120 (1970).

Schiemann: Vom Erlernen unbenannter Anzahlen bei Dohlen. Z. Tierpsychol. **3**, 292 – 347 (1939).

Schlegel, R.: Die Ernährung des Ziegenmelkers (*Caprimulgus europaeus* L.). Beitr. Vogelk. **13**, 145 – 190 (1967).

Schoener, T. W.: Models of optimal size for solitary predators. Am. Naturalist **103**, 277 – 313 (1969 a).

Schoener, Th.: Optimal size and specialization in constant and fluctuating environments: An energy-time approach. Brookhaven Symp. Biol. **22**, 103 – 114 (1969 b).

Schoener, Th.: On the theory of feeding strategies. Am. Rev. Ecol. Syst. **2**, 369 – 404 (1971).

Schreurs, T.: *Lanius collurio* L. und *Lanius senator* L. Ein Beitrag zur Biologie zweier Würgerarten. J. Orn. **84**, 443 – 470 (1936).

Schuh, J., Tietze, F., Schmidt, P.: Beobachtungen zum Aktivitätsverhalten der Wildkatze (*Felis silvestris* Schreber). Hercynia **8**, 102 – 107 (1971).

Schuler, W.: Die Schutzwirkung künstlicher Bates'scher Mimikry abhängig von Modellähnlichkeit und Beuteangebot. Z. Tierpsychol. **36**, 71 – 127 (1974).

Seghers, B. H.: An analysis of geographic variation in the antipredator adaptations of the guppy, *Poecilia reticulata*. Ph. D. thesis, Univ. Brit. Columbia (1973).

Seitz, A.: Die Paarbildung bei *Astatotilapia strigigena* Pfeffer. Z. Tierpsychol. **4**, 40 – 84 (1940).

Selander, R. K.: Sexual dimorphism and differential niche utilization in birds. Condor **68**, 113 – 151 (1966).

Sette, O.: Biology of the Atlantic mackerel (*Scomber scombrus*) of North America. Part II – Migrations and habits. Fisheries Bull. Fish Wildlife Serv. **51**, 251 – 358 (1950).

Shaffer, L. C.: Specializations in the feeding behaviour of gulls and other birds. Ph. D. thesis, Univ. Oxford (1971).

Sheppard, P. M.: A note on non-random mating in the moth *Panaxia dominula* (L.). Heredity **6**, 239 – 241 (1952).

Siebenaler, J. B., Caldwell, D. K.: Cooperation among adult dolphins. J. Mammal. **37**, 126 – 128 (1956).

Siegfried, W. R.: Aspects of the feeding ecology of cattle egrets (*Ardeola ibis*) in South Africa. J. Anim. Ecol. **41**, 71 – 78 (1972).

Simmons, J. A., Wever, E. G., Pylka, J. M.: Periodical cicada: Sound production and hearing. Science **171**, 212 – 213 (1971).

Simons, S., Alcock, J.: Learning and the foraging persistence of white-crowned sparrows *Zonotrichia leucophrys*. Ibis **113**, 477 – 482 (1971).

Sisson, R. F.: Aha! It really works! Nat. Geograph. Mag. **145**, 143 – 147 (1974).

Skoczeń, S.: Gromadzenie zapasów pokarmowych przez niektóre ssaki owadozerne (Insectivora). Przeglad Zoologiczny **14**, 243 – 248 (1970).

Skutch, A. F.: Do tropical birds rear as many young as they can nourish? Ibis **91**, 430 – 455 (1949).

231

Slobodkin, L. B.: How to be a predator. Am. Zoologist **8**, 43 – 51 (1968).

Slobodkin, L. B.: Prudent predation does not require group selection. Am. Naturalist **108**, 665 – 678 (1974).

Smeenk, C.: Ökologische Vergleiche zwischen Waldkauz *Strix aluco* und Waldohreule *Asio otus.* Ardea **60**, 1 – 71 (1972).

Smith, J. N. M.: Studies of the searching behaviour and prey recognition of certain vertebrate predators. Ph. D. thesis, Univ. Oxford (1971).

Smith, J. N. M., Dawkins, R.: The hunting behaviour of individual great tits in relation to spatial variations in their food density. Anim. Behav. **19**, 695 – 706 (1971).

Smith, N. G.: Provoked release of mobbing – a hunting technique of *Micrastur* falcons. Ibis **111**, 241 – 243 (1969).

Smith, S. M.: A study of prey-attack behaviour in young loggerhead shrikes, *Lanius ludovicianus* L. Behav. XLIV, 113 – 141 (1973).

Snow, D. W.: A field study of the bearded bellbird in Trinidad. Ibis **112**, 299 – 329 (1970).

Snyder, N. F. R.: An alarm reaction of aquatic gastropods to intraspecific extract. In: Cornell Univ. Agr. Exp. Sta. Mem. **403**, 1 – 122 (1967).

Snyder, N. F. R., Snyder, H. A.: Defenses of the Florida apple snail *Pomacea paludosa.* Behav. **40**, 175 – 215 (1971).

Soane, I. D., Clarke, B.: Evidence for apostatic selection by predators using olfactory cues. Nature **241**, 62 – 64 (1973).

Soljan, T.: *Blennius galerita* L., poisson amphibien des zones supra-littorale et littorale exposeés de l'Adriatique. Acta Adriat. **2**, 1 – 14 (1932).

Someren, V. G. L. van, Jackson, T. H. W.: Some comments on protective resemblance amongst African Lepidoptera (Rhopalocera). J. Lepidopterists' Soc. **13**, 121 – 150 (1959).

Sommerhoff, G.: Analytical Biology. Oxford Univ. Press 1950.

Southern, H. N.: Tawny owls and their prey. Ibis **96**, 384 – 410 (1954).

Southwood, T. R. E.: Ecological Methods – with Particular Reference to the Study of Insect Populations. London: Methuen and Co. Ltd. 1966.

Spalding, D. J., Lesowski, J.: Winter food of the cougar in south-central British Columbia. J. Wildlife Management **35**, 378 – 381 (1971).

Sparks, J., Soper, T.: Owls. Their Natural and Unnatural History. Newton Abbot: David and Charles Ltd. 1970.

Sparrowe, R. D.: Prey-catching behaviour in the sparrow hawk. J. Wildlife Management **36**, 297 – 308 (1972).

Sperling, G., Budiansky, J., Spivak, J. C., Johnson, M. C.: Extremely rapid visual search. The maximum rate of scanning letters for the presence of a numeral. Science **174**, 307 – 311 (1971).

Springer, C. G., Smith-Vaniz, W. F.: Mimetic relationships involving fishes of the family Blenniidae. Smithson. Contrib. Zool. **112**, 1 – 36 (1972).

Starck, W. A. II., Davis, W. P.: Night habits of fishes of alligator reef, Florida. Ichthyologica **38**, 313 – 356 (1966).

Steiniger, F.: Beiträge zur Soziologie und sonstigen Biologie der Wanderratte. Z. Tierpsychol. **7**, 356 – 379 (1950).

Stower, W. J., Greathead, D. J.: Numerical changes in a population of the desert locust, with special reference to factors responsible for mortality. J. Appl. Ecol. **6**, 203 – 235 (1969).

Sulkava, S.: Zur Nahrungsbiologie des Habichts, *Accipiter gentilis.* Aquilo Ser. Zool. **3**, 1 – 103 (1964).

Sumner, F. B.: Does "protective coloration" protect? Results of some experiments with fishes and birds. Proc. Natl. Acad. Sci. U.S. **20**, 559 – 564 (1934).

Sumner, F. B.: Evidence for the protective value of changeable coloration in fishes. Am. Naturalist LXIX, 245 – 266 (1935).

Sunkel, W.: Der Vogelfang für Wissenschaft und Vogelpflege. Hannover 1927.

232

Svoboda, B.: Die Bedeutung von Farb-, Form- und Geruchsmerkmalen für das Vermeidenlernen von Beuteobjekten bei *Lacerta agilis* L. Dissertation Wien 1969.

Swynnerton, C. F. M.: Experiments and observations on the explanation of form and colouring. J. Linnean Soc. London **33**, 203 – 285 (1919).

Tamisier, A.: Signification du grégarisme diurné et de l'alimentation nocturne des sarcelles d'hiver *Anas crecca crecca* L. Terre Vie **4**, 511 – 562 (1970).

Tartar, V.: The biology of *Stentor*. New York, London: Pergamon Press 1961.

Teleki, G.: The predatory behaviour of wild chimpanzees. Bucknell Univ. Press 1973.

Tenovuo, R., Lemmetyinen, R.: On the breeding ecology of the starling *Sturnus vulgaris* in the archipelago of southwestern Finland. Ornis Fenn. **47**, 159 – 166 (1970).

Teytaud, A. R.: Food habits of the goby, *Ginsburgellus novemlineatus* and the clingfish, *Arcos rubiginosus*, associated with echinoids in the Virgin Islands. Carib. J. Sci. **11**, 41 – 45 (1971).

Thielcke, G.: Zum Beuteverhalten des Raubwürgers (*Lanius excubitor*) und anderer Mäusejäger. Z. Tierpsychol. **13**, 272 – 277 (1956).

Thomas, G.: The influence of encountering a food object on subsequent searching behaviour in *Gasterosteus aculeatus* L. Anim. Behav. **22**, 941 – 952 (1974).

Thomas, K.: Predatory behaviour in two strains of laboratory mice. Psychon. Sci. **15**, 13 – 14 (1969).

Thomas, K.: Predatory behaviour in mice: Strain and sex comparisons. Paper Western Psychol. Ass. Meeting April 1971 a.

Thomas, K.: How predatory behaviour compares with fighting in mice. Paper Animal Behav. Soc. June 1971. Anim. Behav. **19**, 616 (1971 b).

Thomas, K., Fried, M.: Speed of predatory attack in *Mus musculus, Peromyscus californicus*, and *Onichomys torridus*. Paper Am. Soc. Mammalog., Vancouver, B. C., 1971.

Thorpe, W. H.: Learning and Instinct in Animals (2nd ed.). London: Methuen 1958.

Tinbergen, L.: The dynamics of insect and bird populations in pine woods. Arch. Néerl. Zool. **13**, 259 – 473 (1960).

Tinbergen, N.: The Study of Instinct. 2nd Ed. Oxford: Clarendon Press 1952.

Tinbergen, N.: Die Welt der Silbermöwe. Göttingen, Berlin, Frankfurt (Main): Musterschmidt-Verlag 1958.

Tinbergen, N.: Von den Vorratskammern des Rotfuchses (*Vulpes vulpes* L.). Z. Tierpsychol. **22**, 119 – 149 (1965).

Tinbergen, N., and Redaktion von LIFE: Tiere und ihr Verhalten. Time-Life. Int. Nederland N. V. 1966.

Tinbergen, N., Broekhuysen, G. J., Feekes, F., Houghton, J. C. W., Kruuk, H., Szulc, E.: Egg shell removal by the blackheaded gull, *Larus ridibundus* L.: A behaviour component of camouflage. Behav. **19**, 74 – 117 (1962).

Tinbergen, N., Impekoven, M., Franck, D.: An experiment on spacing out as a defense against predation. Behav. **28**, 307 – 321 (1967).

Tolonen, K. E.: Ring-billed gull and laughing gull catch fish by "ploughing" and "skimming". Wils. Bull. **82**, 222 – 223 (1970).

Treat, A. E.: A five-year census of the moth ear mite in Tyringham, Massachusetts. Ecology **39**, 629 – 631 (1958).

Trewawas, E.: Lake Albert fishes of the genus *Haplochromis*. Am. Mag. Nat. Hist. **11**, 435 – 449 (1938).

Trivers, R. L.: The evolution of reciprocal altruism. Quart. Rev. Biol. **46**, 35 – 57 (1971).

Trpis, M.: Development and predatory behaviour of *Toxorhynchites brevipalis* (Diptera: Culicidae) in relation to temperature. Environ. Entomol. **1**, 537 – 546 (1972).

Tugendhat, B.: The normal feeding behavior of the three spined stickleback (*Gasterosteus aculeatus* L.). Behav. **15,** 284 – 318 (1960).

Tullock, G.: The coal tit as a careful shopper. Am. Naturalist 77 – 80 (1969).

Turnbull, A. L.: The search for prey by a web-building spider *Achaeranea tepidariorum* (C. L. Koch) (Araneae, Theridiidae). Can. Entomologist **96,** 568 – 579 (1964).

Turner, E. R. A.: Social feeding in birds. Behav. **24,** 1 – 46 (1965).

Uematsu, T.: Social facilitation in feeding behavior of the guppy. I. Preliminary experiment. II. Experimental analysis of mechanisms. III. Influences of the social facilitation upon the activity respiration and growth. Jap. J. Ecol. **21,** I. 48 – 51, II. 54 – 67, III. 96 –103 (1971).

Uexküll, J. v., Kriszat, G.: Streifzüge durch die Umwelten von Tieren und Menschen. Berlin: Springer Verlag 1934.

Ullrich, B.: Untersuchungen zur Ethologie und Ökologie des Rotkopfwürgers (*Lanius senator*) in Südwestdeutschland im Vergleich zu Raubwürgern (*L. excubitor*), Schwarzstirnwürger (*L. minor*) und Neuntöter (*L. collurio*). Vogelwarte **26,** 1 – 77 (1971).

Verner, J.: Time budget of the male long-billed marsh wren during the breeding season. Condor **67,** 125 – 139 (1965).

Vries, T. de: The Galapagos Hawk. Ph. D. Thesis, Univ. Amsterdam, 1973.

Wallace, A. R.: Darwinism. London: MacMillan & Co. 1889.

Ward, P.: Feeding ecology of the black-faced dioch *Quelea quelea* in Nigeria. Ibis **107,** 173 – 214 (1965).

Ward, P., Zahavi, A.: The importance of certain assemblages of birds as "information-centers" for food-finding. Ibis **115,** 517 – 534 (1973).

Ware, D. M.: Predation by rainbow-trout (*Salmo gairdneri*): The effect of experience. J. Fisheries Res. Board Can. **28,** 1847 – 1852 (1971)

Ware, D. M.: Predation by rainbow trout (*Salmo gairdneri*): The influence of hunger, prey density, and prey size. J. Fisheries Res. Board Can. **29,** 1193 – 1201 (1972).

Waters, V. L.: Food-preference of the nudibranch *Aeolidia papillosa,* and the effect of the defenses of the prey on predation. Veliger **15,** 174 – 192 (1973).

Watkins, J. F., Gehlbach, F. R., Baldridge, R. S.: Ability of the blind snake, *Leptotyphlops dulcis,* to follow pheromone trails of army ants, *Neivamyrmex nigrescens* and *N. opacithorax.* Southwestern Naturalist **12,** 455 – 462 (1967).

Webster, F. A.: Interception performance of echolocating bats in the presence of interference. Animal Sonar Syst., Frascati, **I:** 673 – 713 (1967 a).

Webster, F. A.: Some acoustical differences between bats and men. Proc. Intern. Conf. Sensory Devices for the Blind. London: St. Dunstan's 63 – 87 (1967 b).

Webster, F. A., Griffin, D. R.: The role of flight membranes in insect capture by bats. Anim. Behav. **10,** 332 – 340 (1962).

Wells, M. J.: Factors affecting reactions to *Mysis* by newly hatched *Sepia.* Behav. **13,** 96 – 111 (1958).

Wells, M. J., Buckley, S. K. L.: Snails and trails. Anim. Behav. **20,** 345 – 355 (1972).

Welsh, J. H.: Specific influence of the host in the light responses of parasitic water mites. Biol. Bull. **61,** 497 – 499 (1931).

Welty, J. C.: Experiments in group behaviour of fishes. Physiol. Zool. **7,** 85 – 128 (1934).

Wemmer, Ch.: Impaling behaviour of the loggerhead shrike, *Lanius ludovicianus* Linnaeus, Z. Tierpsychol. **26,** 208 – 224 (1969).

White, C. M.: Prairie falcon displays – accipitrine and circinnine hunting methods. Condor **64,** 439 – 440 (1962).

White, C. M., Weeden, R. B.: Hunting methods of gyrfalcons and behaviour of their prey (Ptarmigan). Condor **68,** 517 – 519 (1966).

Wickler, W.: Über das Verhalten der Blenniiden *Runula* und *Aspidontus* (Pisces, Blenniidae). Z. Tierpsychol. **18,** 421 – 440 (1961).

Wickler, W.: Zum Problem der Signalbildung, am Beispiel der Verhaltensmimikry zwischen *Aspidontus* und *Labroides* (Pisces, Acanthopterygii). Z. Tierpsychol. **20,** 657 – 679 (1963).

234

Wickler, W.: Ein Augen fressender Buntbarsch. Natur. Museum **96**, 311 – 315 (1966).

Wickler, W.: Vergleichende Verhaltensforschung und Phylogenetik. In: Die Evolution der Organismen **I**, 420 – 508, G. Heberer (ed.) Stuttgart: Gustav Fischer Verlag 1967.

Wickler, W.: Mimicry. München: Kindler 1968.

Wilbur, H. M., Collins, J. P.: Ethological aspects of amphibian metamorphosis. Science **182**, 1305 – 1314 (1973).

Williams, G. C.: Adaptation and natural selection. Princeton: Princeton Univ. Press 1966.

Willis, E. O.: A study of the foraging behaviour of two species of ant-tanagers. Auk **77**, 160 – 170 (1960).

Willis, E. O.: The behaviour of the spotted antbirds. Ornithol. Monogr. **10**, Am. Ornithol. Union, 1972.

Wilson, E. O.: The insect societies. Cambridge, Mass.: Belknap Press, Harvard Univ. Press 1972.

Wilson, F.: Adult reproductive behaviour in *Asolcus basalis* (Hymenoptera: Scelionidae). Australian J. Zool. **9**, 737 – 751 (1961).

Windell, J. T.: Rate of digestion in the bluegill sunfish. Invest. Indiana Lakes Streams **7**, 185 – 214 (1966).

Winkler, H.: Beiträge zur Ethologie des Blutspechts (*Dendrocopos syriacus*). Das nicht-reproduktive Verhalten. Z. Tierpsychol. **31**, 300 – 325 (1972).

Winkler, H.: Nahrungserwerb und Konkurrenz des Blutspechts, *Picoides* (*Dendrocopos*) *syriacus*. Oecologia **12**, 193 – 208 (1973).

Wodinsky, J.: Movement as a necessary stimulus of *Octopus* predation. Nature **229**, 493 – 494 (1971).

Wolda, H.: Response decrement in the prey catching activity of *Notonecta glauca* L. (Hemiptera). Arch. Néerl. Zool. **14**, 61 – 89 (1961).

Wood, L. H.: Physiological and ecological aspects of prey selection by the marine gastropod, *Urosalpinx cinerea* (Say). Ph. D. thesis, Cornell University (1965).

Woodring, J. P.: Environmental regulation of andropolymorphism in tyroglyphids (Acari). Proc. 2nd Intern. Congr. Acarol., Akad. Kiadó, Budapest, 433 – 440 (1969).

Yarnall, J. L.: Aspects of the behaviour of *Octopus cyanea* Gray. Anim. Behav. **17**, 747 – 754 (1969).

Young, E. C.: Feeding habits of the South Polar Skua, *Catharacta maccormicki*. Ibis **105**, 301 – 318 (1963).

Yudin, B. S.: Storing of earthworms by Siberian mole is one of the adaptations to the life under Siberian climatic conditions. Izv. Sibirsk. Otd. Akad. Nauk CCCP, Ser. Biol. Nauk **3**, 133 – 137 (1972).

Zafiriou, O., Whittle, K. J., Blumer, M.: Response of *Asterias vulgaris* to bivalves and bivalve tissue extracts. Marine Biol. **13**, 137 – 145 (1972).

Zaret, T. A.: Predator-prey interaction in a tropical lacustrine ecosystem. Ecology, **53**, 248 – 257 (1972).

Zaret, T. M., Rand, A. S.: Competition in tropical stream fishes: Support for the competitive exclusion principle. Ecology **52**, 336 – 342 (1971).

Zbinden, K.: Verhaltensstudien an *Serrasalmus nattereri*. Rev. Suisse Zool. **80**, 521 – 542 (1973).

Zeiss, F.: Death of a great crested grebe chick from choking on a fish. La Mevo **21**, 5 – 7 (1974).

Subject Index

239

Scientific Names of Animals and Plants

Animals

Abdim's stork	*Sphenorhynchus abdimii*
African crocodile	*Crocodilus niloticus*
African fish eagle	*Haliaetus vocifer*
African flycatcher	*Diaphorophya castanea*
African paradise flycatcher	*Tschitrea batesi*
African land snail	*Limicolaria martensiana*
African migratory locust	*Locusta migratoria*
African wild dog	*Lycaon pictus*
Alligator snapping turtle	*Macroclemys temminckii*
American kestrel	*Falco sparverius*
American robin	*Turdus migratorius*
American toad	*Bufo fowleri*
Anchoveta	*Cetengraulis mysticetus*
Angler(fish)	Antennariidae, Lophiiformes
Anole (Am.: lizard)	*Anolis lineatopus;* also *Anolis* sp.
Ant lion	Myrmeleonidae (Larva)
Ant-tanager	*Habia* sp.
Apple snail see Florida apple snail	
Aphid	*Aphis fabae*
Archer fish	*Toxotes jaculatrix*
Arctic loon	*Gavia arctica*
Arctic skua	*Stercorarius parasiticus*
Army ant	*Eciton burchelli, Labidus predator*
Axolotl (Salamander)	*Ambystoma* sp.
Babblers	Timaliidae
Bald eagle	*Haliaetus leucocephalus*
Banded mongoose	*Mungos mungo*
Bank vole	*Clethrionomys glareolus*
Barn owl	*Tyto alba*
Barracuda	*Sphyraena* sp.
Bass see largemouth bass	
Bat	*Myotis lucifugus*
Bearded seal	*Erignathus barbatus*
Bison	*Bison bison*
Black-bellied plover	*Pluvialis squatarola*
Blackbird (European)	*Turdus merula*
Black bear	*Ursus americanus*
Black-breasted buzzard kite	*Hamirostra melanosternum*
Black-capped chickadee	*Parus atricapillus*
Black-headed gull	*Larus ridibundus*
Black mamba	*Dendroaspis polylepis*
Black stork	*Ciconia nigra*
Black vulture	*Necrosyrtes (Aegypius) monachus*
Bleak	*Alburnus alburnus*
Blowfly, bluebottle	*Calliphora erythrocephala*

Bluegill sunfish	*Lepomis gibbosus*
Blue heron	*Butorides sundevalli*
Blue jay	*Cyanocitta cristata*
Blue tit	*Parus caeruleus*
Blue-gray gnatcatcher	*Polioptila caerulea*
Broad-winged hawk	*Buteo platypterus*
Brown bear	*Ursus arctos*
Brown pelican	*Pelecanus occidentalis*
Buffalo	*Syncerus caffer*
Bunting	*Emberiza* sp.
Caracal	*Felis caracal*
Caracara	*Polyborus cheriway*
Caribou	*Rangifer tarandus*
Carmine bee-eater	*Merops nubicus*
Carrion crow, crow	*Corvus corone*
Catfish	*Chaca chaca*
Cattle egret	*Bubulcus ibis*
Chaffinch	*Fringilla coelebs*
Chanting goshawk	*Melierax metabates*
Cheetah	*Acinonyx jubatus*
Chestnut-backed chickadee	*Parus rufescens*
Chimpanzee	*Pan troglodytes*
Cichlid	*Haplochromis livingstonii*
Ciliate	*Woodruffia metabolica*
Clawed frog	*Xenopus laevis*
Cleaning wrasse	*Labroides dimidiatus*
Clingfish	*Lepadichthys lineatus*
Coal tit	*Parus ater*
Cobra	*Naja naja*
Common caracara	*Polyborus (plancus) cheriway*
Common five-lined skink	*Eumeces fasciatus*
Common grackle	*Quiscalus quiscula*
Common lizard	*Lacerta vivipara*
Common loon	*Gavia immer*
Common tern	*Sterna hirundo*
Common toad see European toad	
Common viper	*Vipera berus*
Cooper's hawk	*Accipiter cooperi*
Coot	*Fulica atra*
Cornbeetle	*Alphitobius diaperinus*
Cottonmouth mocassin	*Agkistrodon piscivorus*
Coyote	*Canis latrans*
Crayfish	*Orconectes limosus*
Cricket	*Acheta domesticus*
Crowned eagle	*Stephanoaetus coronatus*
Crow see carrion crow	
Cunner	*Tautogolabrus adspersus*
Cuttlefish	*Sepia officinalis*
Dall sheep	*Ovis dalli*
Darwin's finches	Geospizinae
Deer mouse	*Peromyscus leucopus*
Desert locust	*Schistocerca gregaria*
Digger wasp	*Philanthus bicinctus*
Digger wasp, bee-hunting	*Philanthus triangulum*
Digger wasps	Sphecidae
Double-crested cormorant	*Phalacrocorax auritus*
Dragonfly	*Aeschna cyanea*

244

Drongos	Dicruridae
Dwarf mongoose	*Helogale undulata rufula*
Eagle owl	*Bubo bubo*
Ear mite (moth ear mite)	*Myrmonyssus phalaenodectes*
Eastern kingbird	*Muscivora (Tyrannus) tyrannus*
Eider	*Somateria mollissima*
Egyptian vulture	*Neophron percnopterus*
Eland	*Taurotragus oryx*
Elf owl	*Micrathene whitneyi*
Elk	*Cervus canadensis*
Emperor penguin	*Aptenodytes forsteri*
Eucosmid moth	*Ernarmonia conicolana*
European asp	*Vipera aspis*
European bee-eater	*Merops apiaster*
European blackbird see backbird	
European cuckoo	*Cuculus canorus*
European flycatcher	*Ficedula* sp.
European hare	*Lepus europaeus*
European jay	*Garrulus glandarius*
European minnow	*Phoxinus phoxinus*
European nighthawk	*Caprimulgus europaeus*
European toad	*Bufo bufo*
Fairy bluebirds	Irenidae
Feral pigeon	*Columba livia* var. *domestica*
Field frog	*Rana pipiens*
Field mouse	*Apodemus sylvaticus*
Fishing owl (Tawny f. o.)	*Ketupa flavipes*
Flag cabrilla	*Ephinephelus labriformis*
Flagfish	*Jordanella floridae*
Flamingos	Phoenicopteridae
Flatiron herring	*Harengula thrissina*
Florida alligator	*Alligator mississippiensis*
Florida apple snail	*Pomacea paludosa*
Flukes	Trematoda
Forest falcon see Panamanian f. f.	
Forster's tern	*Sterna forsteri*
Frigate bird	*Fregata* sp.
Gafftopsail pompano	*Trachinotus rhodopus*
Galapagos hawk	*Buteo galapagoensis*
Galapagos short-eared owl	*Asio flammeus galapagoensis*
Galapagos mockingbird	*Nesomimus* sp.
Galapagos woodpecker finch see woodpecker finch	
Garden warbler	*Sylvia borin*
Garter snake	*Thamnophis sirtalis*
Geckos	Gekkonidae
Gemsbok	*Oryx gazella*
Giant gourami	*Helostoma temmincki*
Giant scolopender	*Scolopendra galapagoensis*
Glaucous gull	*Larus hyperboreus*
Glutton	*Gulo gulo*
Goldcrest	*Regulus* sp.
Goatfish	*Mulloidichthys dentatus*
Golden eagle	*Aquila chrysaëtos*
Goldfish	*Carassius auratus*
Goshawk	*Accipiter gentilis*
Great black-back	*Larus marinus*

Great blue heron	*Ardea herodias*
Great bustard	*Otis tarda*
Great grey shrike	*Lanuis excubitor*
Greater horseshoe bat	*Rhinolophus ferrum-equinum*
Greater waxmoth	*Galleria mellonella*
Greater yellowlegs	*Totanus melanoleucus*
Great skua	*Stercorarius skua*
Green heron	*Butorides virescens*
Griffon vulture	*Gyps fulvus*
Grizzly bear	*Ursus arctos horribilis*
Ground hornbill	*Bucorvus abyssinicus*
Ground squirrel	*Spermophilus beechyei*
Grouper	*Sebastodes dimidiatus*, also *Mycteroper-ca*, *Epinephelus*
Guppy	*Poecilia reticulata*
Gyrfalcon	*Falco rusticolus*
Hammerhead stork	*Scopus umbretta*
Hawkmoth	*Erinnyis ello*
Hare	*Lepus capensis*
Hedgehog	*Erinaceus europaeus*
Herring	*Clupea harengus*
Herring gull	*Larus argentatus*
Hobby	*Falco subbuteo*
Honeybee	*Apis mellifica*
Holothurian	*Holothuria difficilis*
Hooded vulture	*Necrosyrtes monachus*
Hornbills	Bucerotidae
Hornbill	*Tropicranus albocristatus*
Housefly	*Musca domestica*
Hyena see Spotted Hyena	
Hymenopteran parasite	*Diaeretiella rapae*
Ichneumoid parasite	*Ephialter laticeps*
Jackal	*Canis* sp.
Jackal (golden or Asiatic)	*Canis aureus*
Jackdaw	*Coleus monedula*
Japanese oyster drill	*Ocinebra japonica*
Kestrel	*Falco tinnunculus*
Killer whale	*Orcinus orca*
Kingfisher	*Alcedo atthis*
Kongoni	*Alcelaphus buselaphus cooki*
Kori bustard	*Ardeotis kori*
Lace monitor	*Varanus varius*
Ladybeetle	*Adalia bipunctata*, *Coccinella septempunctata*, *Propylea quatuor-de-coinpunctata*
Lanner falcon	*Falcon biarmicus*
Lapwing	*Vanellus vanellus*
Largemouth bass	*Micropterus salmoides*
Laughing gull	*Larus atricilla*
Leopard	*Panthera pardus*
Lesser spotted eagle	*Aquila pomarina*
Lesser waxmoth	*Achroea grisella*
Lightfoot crab	*Grapsus grapsus*
Lion	*Panthera leo*
Little spotted cat, Tiger cat, ocelot cat	*Leopardus tigrinus*
Loggerhead shrike	*Lanius ludovicianus*
Long-eared owl	*Asio otus*

Longnose gar	*Lepisosteus osseus*
Long-tailed skua	*Stercorarius longicaudus*
Long-tailed tit	*Aegithalos caudatus*
Mallard	*Anas platyrhynchos*
Mangrove finch	*Cactospiza heliobates*
Marsh harrier	*Circus aeruginosus*
Marsh mongoose	*Herpestes (Atilax) paludinosus*
Meadow pipit	*Anthus pratensis*
Mealworm	*Tenebrio molitor* (larva)
Merlin	*Falco columbarius*
Mink	*Mustela vison*
Minnow	*Notropis atherinoides*
Mite	*Myrmonyssus phalaenodectes*
—, predatory	*Amblyseius largoensis*
—, herbivorous	*Oligonychus punicae*
Mocking birds	*Nesomimus* sp.
Mole	*Talpa europaea*
Mongoose	*Herpestes* sp.; *Herpestes i. ichneumon*
Moose	*Alces alces*
Mosquito	*Toxorhynchites brevipalis*, Culicidae
Mosquito	*Aedes aegypti*, Culicidae
Mosquito fish	*Gambusia affinis*
Mountain lion, puma	*Profelis (Puma) concolor*
Mourning dove	*Zenaidura macroura*
Mule deer	*Odocoileus hemionus*
Mullet	*Mullus* sp.
Murray eels	Muraenidae
Nightcrawler	*Lumbricus terrestris*
Nightjars	Caprimulgidae
Nile crocodile	*Crocodilus niloticus*
Norway rat	*Epimys norvegicus*
Nudibranch	*Aeolidia papillosa*
Nuthatch	*Sitta europaea*
Ocelot cat (see also Little spotted cat)	*Leopardus tigrinus*
Octopus	*Octopus vulgaris*
Olive baboon	*Papio anubis*
Orb-web spider	*Argiope argentata*
Osprey	*Pandion haliaetus*
Owls	*Strigiformes*
Oyster	*Ostrea* sp.
Oystercatcher	*Haematopus ostralegus*
Pacific cornetfish	Aulostomidae
Parrot fish	Scaridae
Perch	*Perca fluviatilis*
Peregrine (falcon)	*Falco peregrinus*
Periodic cicada	*Magicicada* sp.
Pheasant	*Phasianus colchicus*
Pied flycatcher	*Ficedula hypoleuca*
Pigmy chamaeleon	*Microsaurus pumilus*
Pigmy owl	*Glaucidium passerinum*
Pike	*Esox lucius*
Panamanian forest falcons	*Micrastur mirandollei, M. semitorquatus*
Pine marten	*Martes martes*
Pipe fish	*Aulostomus maculatus*
Piranha	*Serrasalmus nattereri, S. rhombus*
Plover	*Charadrius* sp.

Polar bear	*Ursus maritimus*
Polecat	*Putorius (Mustela) putorius*
Pollock	*Pollachius virens*
Pony fish	*Leiognathus equulus*
Porpoise	*Phocaena vomerina*
Prairie falcon	*Falco mexicanus*
Prawn	*Leander* sp.
Praying mantis	*Stagmatoptera biocellata,*
	Phyllovates chloropaea,
	Hierodula crassa
Pronghorn antelope	*Antilocapra americana*
Ptarmigan	*Lagopus mutus, L. lagopus*
Puffer fish	Tetraodontidae
Puffin	*Fratercula arctica*
Puma see mountain lion	
Raccoon	*Procyon lotor*
Rainbow trout	*Salmo gairdneri*
Rat snake	*Elaphe obsoleta*
Raven	*Corvus corax*
Red-eared turtle	*Pseudemys scripta*
Redfooted booby	*Sula sula*
Red fox	*Vulpes vulpes*
Redbacked shrike	*Lanius collurio*
Red squirrel	*Sciurus vulgaris*
Red-tailed hawk	*Buteo jamaicensis*
Red-throated loon	*Gavia stellata*
Red-winged blackbird	*Agelaius phoeniceus*
Redworm	*Eisenia foetida*
Ring-billed gull	*Larus delawarensis*
Ringed seal	*Pusa hispida*
Ring ouzel	*Turdus torquatus*
Rock dove	*Columba livia*
Roe deer	*Capreolus capreolus*
Rook	*Corvus frugilegus*
Roseate tern	*Sterna dougallii*
Royal tern	*Sterna maxima*
Rudd	*Scardinius erythrophthalmus*
Rufous-naped tamarin	*Saguinus geoffroyi*
Salamander (axolotl)	*Ambystoma gracile*
Sally light-foot crab	*Grapsus grapsus*
Salticid spider	*Epiblemum scenicum*
Sand-eel	*Ammodytes* sp.
Sand lizard	*Lacerta agilis*
Sandwich tern	*Sterna sandvicensis*
Sawfly	*Neodiprion sertifer*
Sea lamprey	*Petromyzon marinus*
Sea lion	*Zalophus wollebaeki*
Sea snake	*Pelamis platurus*
Sea urchin	*Echinometra lucunter*
Señorita	*Oxyjulis californica*
Serval	*Felis serval*
Shama thrush	*Copsychus malabaricus*
Shore crab	*Carcinus maenas*
Short-eared owl see Galapagos short-eared owl	
Shrew	*Sorex cinereus*
Silversides	*Atherina* sp.

248

Silverside	*Menidia menidia*
Skimmer	*Rhynchops nigra*
Skink	*Eumeces* sp.
Skua	*Stercorarius* sp.
Slaty-backed forest falcon	*Micrastur mirandollei*
Slaty-collared forest falcon	*Micrastur semitorquatus*
Slave-making ants	*Formica subintegra*
Smooth green snake	*Opheodrys vernalis blanchardi*
Snail kite	*Rosthramus sociabilis*
Snapping turtle	*Chelydra serpentina*
Snowy egret	*Leucophoyx thula*
Spadefoot toad	*Scaphiopus bombifrons*
Sparrowhawk	*Accipiter nisus*
Spotted flycatcher	*Muscicapa striata*
Spotted hyena	*Crocuta crocuta*
Squid	*Loligo vulgaris*
Squirrel (also Red Squirrel)	*Sciurus vulgaris*
Starling	*Sturnus vulgaris*
Stick insect	Orthoptera, Phasmatodea, Dictyoptera
Stick insects	*Metriotes diocles*
Stoat	*Mustela erminea*
Sunfish	*Lepomis megalotis*
Storks	Ciconiidae
Swallows	Hirundinidae
Swallow-tailed gull	*Creagrus furcatus*
Syrian woodpecker	*Dendrocopos syriacus*
Tawny fish owl (also Fishing owl)	*Ketupa flavipes*
Tawny owl	*Stric aluco*
Teal	*Anas crecca*
Tench	*Tinca tinca*
Thick-billed nutcracker	*Nucifraga caryocatactes*
Thomson's gazelle	*Gazella thomsoni*
Three-spined stickleback	*Gasterosteus aculeatus*
Thrush	*Turdus* sp.
Tiger	*Neofelis tigris*
Toucans	Rhamphastidae
Towhee	*Pipilo* sp.
Trigger fish	*Balistes fuscus*
Trogons	Trogonidae
Trout	*Salmo* sp.
Tyrant Flycatcher	Tyranni
Vampire bat	*Desmodus rotundus*
Viper	*Vipera xanthina palaestinae*
Warbler finch	*Certhidea olivacea*
Water snake	*Natrix sipedon*
Water stick insect	*Ranatra linearis*
Wattled starling	*Creatophora carunculata*
Waxmouth, Greater; Lesser	*Galleria mellonella; Achroea grisella*
White pelican	*Pelecanus onocrotalus*
White stork	*Ciconia ciconia*
White-tailed deer	*Odocoileus virginianus*
Whitethroat	*Sylvia communis*
White-throated sparrow	*Zonotrichia albicollis*
Wild dog	*Lycaon pictus*
Wildebeest	*Connochaetus taurinus*
Willow grouse	*Lagopus lagopus*
Wolf	*Canis lupus*

Woodcock	*Scolopax rusticola*
Wood ibis, woodstork	*Mycteria americana*
Woodlouse	*Armadillidium* sp.
Woodpecker finch, Galapagos w. f.	*Cactospiza pallida*
Woodpigeon	*Columba palumbus*
Worm (Oligochaeta)	*Enchytraeus* sp.
Wrasse	*Halichoeres nicholsi*
Wrasse (cleaning)	*Labroides dimidiatus*
Yellow-bellied sea snake	*Pelamis platurus*
Yellow-bellied racer	*Coluber constrictor mormon*
Yellow perch	*Perca flavescens*
Zebra, Burchell's	*Equus burchelli*
Zebra danio	*Brachydanio rerio*
Zebra fish	*Pterois volitans*

Plants

Avogado	*Persea indica*
Dove weed	*Eremocarpus setigerus*
Hazel	*Corylus avellana*
Oak	*Quercus* sp.

Zoophysiology and Ecology

Managing Editor: D. S. Farner
Editors: W. S. Hoar, J. Jacobs,
H. Langer, M. Lindauer

Vol. 1: P. J. Bentley
Endocrines and Osmoregulation
A Comparative Account of the
Regulation of Water and Salt
in Vertebrates
29 figures. XVI, 300 pages. 1971

Vol. 2: L. Irving
Arctic Life of Birds and Mammals
Including Man
59 figures. XI, 192 pages. 1972

Vol. 3: A. E. Needham
The Signification of Zoochromes
54 figures. XX, 429 pages. 1974

Vol. 4/5: A. C. Neville
Biology of the Arthropod Cuticle
233 figures. XVI, 448 pages. 1975

Vol. 6: K. Schmidt-Koenig
Migration and Homing in Animals
64 figures. XII, 99 pages. 1975

Springer-Verlag
Berlin
Heidelberg
New York

Behavioral Ecology and Sociobiology

Managing Editor:
Hubert Markl, Fachbereich Biologie, Universität, D-775 Konstanz

Editors:
John H. Crook, Bristol; Bert Hölldobler, Cambridge, Mass.;
Hans Kummer, Zurich; Edward O. Wilson, Cambridge, Mass.

Based on quantitative studies, **Behavioral Ecology and Sociobiology** publishes original contributions and short communications dealing with the analysis of animal behavior on both the individual and population level. The functions, mechanisms, and evolution of ecological adaptations of behavior are given special emphasis.

Aspects of particular interest are:

Orientation in space and time. Communication and all other forms of social and interspecific behavioral interaction, including predatory and antipredatory behavior. Origins and mechanisms of behavioral preferences and aversions, e.g., with respect to food, locality, and social partners. Behavioral mechanisms of competition and resource partitioning. Population physiology. Evolutionary theory of social behavior.

Volume 1 (4 issues) will appear in 1976.
For a sample copy as well as subscription information please address:

Springer-Verlag
Werbeabteilung 4021
Heidelberger Platz 3
D-1000 Berlin 33
or
Springer-Verlag
New York Inc.
Promotion Department
175 Fifth Avenue
New York, N.Y. 10010

Springer-Verlag
Berlin
Heidelberg
New York